AF885916

Springer Theses

Recognizing Outstanding Ph.D. Research

Aims and Scope

The series "Springer Theses" brings together a selection of the very best Ph.D. theses from around the world and across the physical sciences. Nominated and endorsed by two recognized specialists, each published volume has been selected for its scientific excellence and the high impact of its contents for the pertinent field of research. For greater accessibility to non-specialists, the published versions include an extended introduction, as well as a foreword by the student's supervisor explaining the special relevance of the work for the field. As a whole, the series will provide a valuable resource both for newcomers to the research fields described, and for other scientists seeking detailed background information on special questions. Finally, it provides an accredited documentation of the valuable contributions made by today's younger generation of scientists.

Theses may be nominated for publication in this series by heads of department at internationally leading universities or institutes and should fulfill all of the following criteria

- They must be written in good English.
- The topic should fall within the confines of Chemistry, Physics, Earth Sciences, Engineering and related interdisciplinary fields such as Materials, Nanoscience, Chemical Engineering, Complex Systems and Biophysics.
- The work reported in the thesis must represent a significant scientific advance.
- If the thesis includes previously published material, permission to reproduce this must be gained from the respective copyright holder (a maximum 30% of the thesis should be a verbatim reproduction from the author's previous publications).
- They must have been examined and passed during the 12 months prior to nomination.
- Each thesis should include a foreword by the supervisor outlining the significance of its content.
- The theses should have a clearly defined structure including an introduction accessible to new PhD students and scientists not expert in the relevant field.

Indexed by zbMATH.

More information about this series at http://www.springer.com/series/8790

Ryosuke Yanagihara

Distribution of Energy Momentum Tensor around Static Charges in Lattice Simulations and an Effective Model

Doctoral Thesis accepted by
Osaka University, Osaka, Japan

Springer

Author
Dr. Ryosuke Yanagihara
Department of Physics
Graduate School of Science
Osaka University
Toyonaka, Osaka, Japan

Supervisors
Prof. Masayuki Asakawa
Department of Physics
Graduate School of Science
Osaka University
Toyonaka, Osaka, Japan

Prof. Masakiyo Kitazawa
Department of Physics
Graduate School of Science
Osaka University
Toyonaka, Osaka, Japan

ISSN 2190-5053 ISSN 2190-5061 (electronic)
Springer Theses
ISBN 978-981-16-6233-1 ISBN 978-981-16-6234-8 (eBook)
https://doi.org/10.1007/978-981-16-6234-8

This Springer imprint is published by the registered company Springer Nature Singapore Pte Ltd.
The registered company address is: 152 Beach Road, #21-01/04 Gateway East, Singapore 189721, Singapore

Supervisors' Foreword

It is a great pleasure for us to introduce Dr. Ryosuke Yanagihara's thesis work, accepted for publication within Springer Theses and awarded with a prize for an outstanding original work. He is the first graduate student awarded this prize in our group. He graduated from Department of Physics at Osaka University *summa cum laude*. After his graduation, he continued his study at the graduate school and was awarded Ph.D. as of March 2021.

During his Ph.D. study, he mainly investigated energy momentum tensor in QCD at zero and finite temperatures. Because of its non-perturbative nature, it is hard to understand QCD quantitatively. Lattice gauge theory is among the most reliable and systematic schemes to handle the non-perturbative feature of QCD. It discretizes space–time. As a result, it is not trivial to define energy momentum tensor, which is the generator of translation in space–time, on the lattice. This problem has been recognized since its birth and has been attacked intensively. One of solutions, gradient-flow method, has recently been established. In the process of establishment, contributions by Lüscher, Weisz, and Suzuki were crucial; they proposed the method, established renormalizability of the theory, and worked out its perturbative expression first. After the completion of the theoretical framework, it was high time that QCD energy momentum tensor should be investigated numerically on the lattice. Dr. Yanagihara was in the right place at the right time. However, his efforts should not be underestimated. The numerical investigation of energy momentum tensor density on the lattice is not that simple. Two non-commutative limits need to be taken. One is a familiar one, continuum limit. The other is the one newly introduced in gradient-flow method, zero t limit with t the fictitious fifth coordinate called flow time. As described in the thesis, the extrapolations should be carried out with a lot of care. In addition, they are fights with noise. After long careful and complicated analyses, he obtained the distribution of the stress tensor between a heavy quark and a heavy anti-quark at zero temperature and the distribution around a heavy quark above the deconfinement temperature for the first time. The results were sufficiently quantitative to be compared with, *e.g.*, perturbative results. We note that energy momentum tensor density is a quantity that is gauge invariant and whose meaning is physically clear, unlike color electric fields, action density, and so forth.

As explained above, the thesis is at the very forefront of the trials to understand QCD microscopically. The readers will agree after reading through the thesis and looking at quite a few of impressive figures. We congratulate Dr. Yanagihara on his outstanding work.

Toyonaka, Osaka, Japan
May 2021

Masayuki Asakawa

In Yang–Mills (YM) gauge theory, the fundamental theory of elementary particles, static quarks are important probes for investigating non-perturbative aspects of the theory. An example is the flux tube, which is formed between a static quark and an anti-quark. Properties of the flux tube are related to fundamental problems in YM theory, such as the confinement of quarks, and have been studied actively over the last few decades.

In this thesis, Dr. Yanagihara performs extensive studies on the static-quark systems. In particular, he has succeeded in performing the first numerical simulation of the distribution of energy momentum tensor around the flux tube in SU(3) YM theory by applying the state-of-the-art numerical technique based on the gradient flow. Since the energy momentum tensor is one of the most fundamental quantities in physics, this establishment opens up various future studies for revealing non-trivial aspects of the strong interaction. In fact, he has conducted one of such studies by himself by studying the energy momentum tensor around the flux tube in a theoretical model, which is the other important constituent of this thesis. Moreover, his numerical analyses are also extended to a single static-quark system at nonzero temperature. This thesis aggregates these outstanding achievements with a well-organized general review.

The gradient flow, which is the key technology of this thesis, had been first introduced for the numerical analysis of the energy momentum tensor in 2014 by our group. Dr. Yanagihara had joined this collaboration in 2016, and after that, he has been at the forefront of the development. The gradient flow is a powerful technique, but at the same time, it requires complex and careful numerical analyses before obtaining physical results. These analyses would not be achieved unless his outstanding contribution. Moreover, his accomplishments were not limited to numerical aspects, but he was also a major contributor to mathematical calculations and theoretical studies in effective models.

As his advisor and a collaborator who has watched him most closely for more than four years, I am very pleased to see the doctoral thesis of Dr. Yanagihara being published as a part of the Springer Theses. I would like to congratulate Dr. Yanagihara on the Springer Theses Award.

Toyonaka, Osaka, Japan
May 2021

Masakiyo Kitazawa

Preface

The energy momentum tensor (EMT) is one of the most fundamental observables in physics. Recently, a novel method to define EMT on the basis of the gradient-flow formalism has been proposed. It has then turned out that the EMT operator via the gradient flow can be employed even on the lattice, which leads to non-perturbative computations. This approach has been successfully applied to the analyses on the thermodynamic quantities.

In this dissertation, we study the spatial EMT distributions around static charges via the gradient flow in the lattice simulations based on the SU(3) Yang–Mills (YM) theory. We employ static charges as probes to explore complex quantum systems. EMT then characterizes the response of the vacuum as well as the hot medium under the existence of the charges, which significantly provides profound and novel insights into the non-perturbative phenomena. In addition, we study the EMT distribution around the magnetic vortex in the Abelian-Higgs (AH) model and compare it with the lattice result.

We first investigate the EMT distribution around a static single quark in the deconfined phase of the SU(3) YM theory. The resultant EMT distribution after the continuum limit shows a sharp contrast to the prediction in the classical Maxwell theory. We then extract the strong coupling constant from the distribution at short distance by comparing the lattice result with the perturbative behavior. We also find that the result also indicates the thermal screening at long distance.

Next we focus on the vacuum of the SU(3) YM gauge theory. The spatial distribution of EMT in the static-quark–anti-quark ($Q\bar{Q}$) system is analyzed on the lattice. The resultant EMT distribution reveals the formation of the flux tube in a gauge-invariant manner. We find remarkable non-Abelian features by studying the transverse structure quantitatively. We also perform a consistency check of our analysis. We confirm the agreement between the force defined from the $Q\bar{Q}$ potential and the force derived from the integral of the stress tensor.

In order to further investigate the local structure of the flux tube, we study the EMT distribution in the $Q\bar{Q}$ system based on the conservation law of EMT and the AH model. Compared with the EMT distribution around $Q\bar{Q}$ in the SU(3) YM theory measured on the lattice in the above setup, the momentum conservation suggests the importance of boundaries corresponding to $Q\bar{Q}$ charges. We then investigate the EMT distribution around the magnetic vortex between a monopole and an anti-monopole with/without boundaries in the AH model and compare it with the lattice results based on the dual superconductor picture. We find that a wide range of parameters in the AH model is excluded since the lattice results cannot be reproduced.

Toyonaka, Osaka, Japan
June 2021

Ryosuke Yanagihara

Acknowledgements

First of all, I wish to sincerely express my gratitude to my supervisor Professor Masayuki Asakawa for supporting my Ph.D. course generously and providing me insightful advices on my researches. I am deeply grateful to Professor Masakiyo Kitazawa for his enthusiastic guidance and persistent help throughout my Ph.D. course.

This Ph.D. dissertation is largely composed of the collaborating works in FlowQCD collaboration. I would like to thank members in FlowQCD collaboration for giving me a lot of chances of interesting communications. I owe my gratitude to Professor Tetsuo Hatsuda for his illuminating comments and suggestions. I would like to thank Dr. Takumi Iritani for having many discussions with me.

The members in Nuclear Theory Group in Osaka University have been supported my researching throughout my Ph.D. course. I would like to thank Professor Yukinao Akamatsu for fruitful discussions and his constructive comments. I also thank all my colleagues in Nuclear Theory Group in Osaka University for their friendships and fruitful communications.

Finally, I would like to express my deep gratitude to my family for their continuous support during my whole life. Without their encouragement, I could not have completed my Ph.D. course.

Parts of this thesis have been published in the following journal articles:

- R. Yanagihara, T. Iritani, M. Kitazawa, M. Asakawa and T. Hatsuda, "Distribution of stress tensor around static quark--anti-quark from Yang-Mills gradient flow," Physics Letters B **789** (2019) 210, doi:10.1016/j.physletb.2018.09.067

  [arXiv:1803.05656 [hep-lat]].
- R. Yanagihara and M. Kitazawa, "A study of stress-tensor distribution around the flux tube in the Abelian-Higgs model," Progress of Theoretical and Experimental Physics **2019**, Issue 9, September 2019, 093B02, doi:10.1093/ptep/ptz093

  [arXiv: 1905.10056 [hep-ph]];
  [Erratum: Progress of Theoretical and Experimental Physics **2020**, Issue 7, July 2020, 079201

- R. Yanagihara, M. Kitazawa, M. Asakawa and T. Hatsuda, "Distribution of energy-momentum tensor around a static quark in the deconfined phase of SU(3) Yang-Mills theory," Physical Review D **102** (2020) 114522, doi: 10.1103/PhysRevD.102.114522 [arXiv:2010.13465 [hep-lat]].

Contents

1 Introduction 1
1.1 Quantum Chromodynamics 3
1.1.1 Asymptotic Freedom 5
1.1.2 Chiral Symmetry 6
1.1.3 Confinement 7
1.2 Flux Tube 8
1.2.1 Dual Abelian-Higgs Model 8
1.2.2 Dynamics of Strings 9
1.2.3 Lattice Approach 10
1.3 Phase Transition and Exploration of Deconfined Phase 12
1.4 Purpose 15
1.4.1 Lattice Study 16
1.4.2 Effective Model 17
1.5 Organization 17
References 19

2 Energy Momentum Tensor 23
2.1 Derivation and Properties of Energy Momentum Tensor 24
2.1.1 Derivation and Properties 24
2.1.2 Physical Meanings 25
2.1.3 Example: Energy Momentum Tensor in Pure Gauge Theory 27
2.2 Energy Momentum Tensor in Maxwell Theory 28
2.2.1 Maxwell Equations and Classical System 28
2.2.2 Energy Conservation 29
2.2.3 Momentum Conservation 31
2.2.4 Maxwell Stress and Coulomb Force 32
2.2.5 Stress Distribution on Source Plane 34
References 35

3 Lattice Field Theory 37
3.1 Feynman Path Integral 38
3.2 Lattice Regularization 39
3.2.1 Simple Example 40
3.2.2 Free Boson Field 40
3.2.3 Free Fermion Field 43
3.3 Lattice Gauge Theory 44
3.3.1 Gauge Principle and Gauge Field 44
3.3.2 Link Variable 45
3.3.3 Path Integral 48
3.3.4 Non-Abelian Gauge Theory on the Lattice 49
3.4 Gauge-Invariant Quantity on the Lattice 50
3.4.1 Wilson Loop 50
3.4.2 Polyakov Loop 51
References 52

4 Yang–Mills Gradient Flow and Energy Momentum Tensor 55
4.1 Quantum Theory of Yang–Mills Theory 56
4.1.1 Action 56
4.1.2 Quantization 57
4.1.3 One-Loop Divergences 58
4.1.4 Renormalization 61
4.2 Introduction to Gradient Flow 63
4.2.1 Idea of Yang–Mills Gradient Flow 63
4.2.2 Perturbative Expansion of Gradient Flow 65
4.2.3 Example: Two-Point Function of Flowed Gauge Field 67
4.3 Proof of Renormalizability of Gradient Flow 70
4.3.1 $(D+1)$-Dimensional Field Theory 70
4.3.2 BRS Invariance and WT Identity of $(D+1)$-Dimensional System 72
4.3.3 Proof of Renormalizability 75
4.4 Application to Energy Momentum Tensor 79
4.4.1 Small Flow Time Expansion 80
4.4.2 Renormalization Group Equation and Expansion Coefficients 82
4.4.3 Energy Momentum Tensor on the Lattice 83
4.5 Comments on Fermions 84
References 86

5 Distribution of Energy Momentum Tensor around Single Static Quark in Deconfined Phase of SU(3) Yang–Mills Theory 89
5.1 EMT around a Static Quark in SU(3) Yang–Mills Theory on the Lattice 90
5.1.1 Correlation 90
5.1.2 Spherical Coordinate 91

5.2 Lattice Setup 92
5.2.1 Gauge Configurations 92
5.2.2 Discretization Effect 94
5.2.3 Double Extrapolation 94
5.3 Results of EMT Distributions 96
5.3.1 Channel Dependence 97
5.3.2 Temperature Dependence 97
References 100

6 Distribution of Energy Momentum Tensor around Static Quark–Anti-Quark in Vacuum of SU(3) Yang–Mills Theory 103
6.1 Wilson Loop and $Q\bar{Q}$ Potential 104
6.1.1 $Q\bar{Q}$ Potential 105
6.1.2 Determination of N_{APE} 107
6.2 Distribution of Energy Momentum Tensor around $Q\bar{Q}$ 110
6.2.1 Correlation 110
6.2.2 Cylindrical Coordinate 111
6.3 Setup 111
6.4 EMT Distribution on Source Plane 113
6.5 EMT Distribution on Mid-Plane 114
6.5.1 Double Extrapolation 114
6.5.2 EMT Distribution on Mid-Plane 116
References 119

7 Distribution of Energy Momentum Tensor around Magnetic Vortex in Abelian-Higgs Model 121
7.1 Stress Tensor and Momentum Conservation in Cylindrical Coordinate 122
7.2 Abelian-Higgs Model 123
7.2.1 Model 124
7.2.2 Energy Momentum Tensor in Cylindrical Coordinate System 125
7.3 Magnetic Vortex 127
7.3.1 Magnetic Vortex with Finite Length 127
7.3.2 Infinitely Long Vortex 128
7.3.3 Physical Units 130
7.3.4 Details of Numerical Analysis 131
7.4 Numerical Results 131
7.4.1 Infinitely Long Vortex 132
7.4.2 Finite-Length Flux Tube 134
References 137

8 Summary and Outlook 139
References 141

Appendix A: Improved Estimator ... 143

Appendix B: Tree-Level Improvement of Lattice Observables ... 145

Appendix C: Leading-Order Perturbative Analysis of EMT around Single Static Charge ... 149

Appendix D: Smearing Method ... 153

Appendix E: Analytic Properties of Abelian-Higgs Model ... 157

Chapter 1
Introduction

In the history of the particle physics, physicists have sought for the most elementary building blocks of all matters and have attempted to reveal their physical properties. They have in turn encountered smaller and smaller ingredients of matters and have afterward found that these constituents are, in most cases, composite ones.

All matters around us are composed of atoms. At the beginning of the twentieth century, it was considered that the atom is the most elementary constituent of all matters. Rutherford's experiment in 1911 made a stir in this situation [1]. Before then in 1909, under his guidance, Geiger and Marsden had measured the angular distribution of scattered α particles[1] from a radium atom toward a gold target [2]. From the existence of large-angle scatterings of the α particles, Rutherford found that a dense and positive-charged nucleus, whose radius is approximately one hundred-thousandth as small as that of a typical atom, resides at the center of the atom. It is today understood that the atom is composed of the nucleus at the center surrounded by the negative-charged electron, which is bound via the Coulomb interaction. While the electron is considered to be an elementary particle in the present Standard Model, the nucleus is composite of the smaller nucleons, the proton, and the neutron. Since Rutherford discovered the proton in 1918 [3] and Chadwick did the neutron in 1932 [4], this picture has been established undoubtedly.

There still remained a question of how nuclei exist stably despite the Coulomb repulsive force among protons. In 1935, Yukawa answered the question [5]. In his theory, the strong interaction acts among nucleons, where the newly introduced particle, called the meson, mediates the interaction. In fact, Powell and his collaborators discovered the π meson from the cosmic rays and demonstrated the Yukawa theory in 1947 [6].[2]

---

[1] The α particle is the nucleus of the helium atom.

[2] It was, at the same time, confirmed that the mass of the π meson is approximately equal to 140 MeV. In 1937, Anderson and Neddermeyer discovered the particle, whose mass is approximately equal to that of the meson which Yukawa had predicted. This particle was, however, found insensitive to the strong interaction. This particle is today known as the muon.

R. Yanagihara, *Distribution of Energy Momentum Tensor around Static Charges in Lattice Simulations and an Effective Model*, Springer Theses,
https://doi.org/10.1007/978-981-16-6234-8_1

Table 1.1 Properties of quarks [11]. The mass is determined in the modified minimal subtraction ($\overline{\text{MS}}$) scheme

Flavor	Mass	Charge	Isospin	Spin/Parity (J^P)
u (Up)	$2.16^{+0.49}_{-0.26}$ MeV	$2/3$	$1/2$	$(1/2)^+$
d (Down)	$4.67^{+0.48}_{-0.17}$ MeV	$-1/3$	$-1/2$	$(1/2)^+$
s (Strange)	93^{+11}_{-5} MeV	$-1/3$	0	$(1/2)^+$
c (Charm)	1.27 ± 0.02 GeV	$2/3$	0	$(1/2)^+$
b (Bottom)	$4.18^{+0.03}_{-0.02}$ GeV	$-1/3$	0	$(1/2)^+$
t (Top)	172.76 ± 0.30 GeV	$2/3$	0	$(1/2)^+$

Until the middle of the twentieth century, the above picture sufficiently explains, to some extent, the phenomena in the atomic and nuclear physics. However, the prosperity in high energy experiments with large accelerators played a role of raising a question on this concise description. In the 1950s and 1960s, these experiments produced more than a hundred particles, so that it became clear that the proton and the neutron as well as the pion are just members of a particle family, called the hadron.[3] The hadrons are classified into the baryons and the mesons according to whether they are fermions or bosons. The former is fermion with a half-odd spin, and the latter is boson with an integer spin. For instance, the proton and the neutron are baryons, while the π meson belongs to a family of mesons.

It has been revealed that even hadrons are not elementary particles but composite ones from, for example, deep inelastic scattering experiments [7, 8]. From the theoretical viewpoint as well, for the systematic classification of too many hadrons, Gell-Mann [9] and Zweig [10] independently proposed the quark model. The quark model is an idea that hadrons are composed of one-step more elementary degrees of freedom. Gell-Mann referred to this elementary particle as the quark.[4] From the modern viewpoint, hadrons are composed of quarks and anti-quarks with gluons, which mediate the strong interaction. In the Standard Model, they all are considered to be elementary particles. In particular, baryons are composite of three valence quarks and mesons are of a valence quark and an anti-quark. Note that, in the quark model, quarks carry fractional charges while hadrons possess integer charges.

Today, it is understood that there are six kinds of quarks and each quark is labeled by the specific flavor. Shown in Table 1.1 is the list of all the flavors with their physical properties [11]: From the left, shown are the flavor, mass, charge, isospin, and spin/parity (J^P). The u and d quarks are referred to as the light quarks. Their masses, about a few MeV, are significantly lighter than the typical energy scale of the strong interaction that is called Λ_{QCD}. The s quark is heavier than the light quarks. Nevertheless, since the mass of the s quark is smaller than Λ_{QCD}, the approximation

[3] More than a hundred hadrons are often called 'Hadron Zoo.'

[4] Although Zweig named the particle *ace*, the name *quark* becomes nowadays popular.

of the flavor SU(3) symmetry, where the masses of the u, d, and s quarks degenerate, can be justified for some phenomena.

In contrast, the c, b, and t quarks are referred to as the heavy quarks. The existence of the c quark was theoretically predicted by Glashow, Iliopoulos, and Maiani [12]. The independent experiments at both Stanford Linear Accelerator Center [13] and Brookhaven National Laboratory [14] in 1974 revealed the existence of the J/ψ particle, which is the bound state of c and $\bar{c}$. The existence of the b and t quarks is also theoretically predicted by Kobayashi and Maskawa in 1973 in order to explain the CP violation of the K meson [15]. In 1977, Υ particle, which is the bound state of b and $\bar{b}$, was discovered at Fermi National Accelerator Laboratory [16]. In 1995, also at Fermilab, the existence of t quark was confirmed through the CDF [17] and D0 [18] experiments.

Although the quark model seems to be a highly organized system of classification, a crucial criticism on the model had been pointed out since its proposal. In order to demonstrate the problem, let us consider the Δ^{++} baryon, whose spin is $S = 3/2$. Since Δ^{++} is composed of three u quarks and has the orbital angular momentum $L = 0$, the spatial wave function is symmetric. The total angular momentum $J = 3/2$ requires the spin of three u quarks to be all parallel, hence the spin wave function be also symmetric. However, this fact contradicts the Pauli principle. In 1965, Han and Nambu solved the question with the introduction of the new degree of freedom for quarks, called the color [19]. Since the color wave function is constructed to be antisymmetric with respect to the interchange of particles, the total wave function becomes antisymmetric. This does not conflict with the Pauli principle. Although the color degree of freedom was, at first, introduced for a theoretical reason, the existence of the color was later confirmed experimentally [20]. In the following, we recapitulate the theory of the strong interaction, which succeeds to the above foundations.

1.1 Quantum Chromodynamics

All the physical phenomena in nature occur via only four fundamental interactions: gravitational, electromagnetic, strong, and weak interactions. The latter three interactions of elementary particles other than the gravity are summarized in the Standard Model.[5] Among these interactions, we particularly focus on the dynamics of the strong interaction carried by quarks and gluons, which is well described via quantum chromodynamics (QCD). In this section, we introduce QCD and discuss its unique properties.

[5] Although the Standard Model is almost consistent with the experimental results so far, there exist some phenomena beyond the Standard Model, such as the fine-tuning problem on the Higgs particle, the strong CP problem, the neutrino oscillation, and so on. For the purpose of the exploration beyond the Standard Model, the grand unified theory and super-string theory are discussed from the theoretical aspects, while the experiments at extremely high energy are in progress.

The theoretically most important building block of the Standard Model is the Yang–Mills (YM) theory proposed in 1954 by Yang and Mills [21], and in 1956 by Uchiyama [22]. The YM theory favors the invariance under the local gauge transformation. At the beginning of the proposal by Yang and Mills, in order to represent the isospin SU(2) symmetry, they extended the well-known Abelian gauge theory to the non-Abelian one [21]. Although this attempt, in this sense, aimed at the description of the strong force, the SU(2) YM theory in favor of the isospin symmetry between the proton and the neutron did not match the experimental results. Nevertheless, the direction along the SU(2) YM theory succeeded in the Weinberg–Salam theory, which unifies the electromagnetic and weak interactions [23–25].

Now, we turn to QCD based on the SU(3) YM gauge theory in the D-dimensional space. QCD respects the internal color SU(3) symmetry, whose action is written down as

$$S = S_{\mathrm{mat}} + S_{\mathrm{G}}, \tag{1.1}$$

$$S_{\mathrm{mat}} = \int d^D x \sum_{f-\mathrm{flavors}} \bar{\psi}^{\alpha}_{f}(x)(\not{D}_{\alpha\beta} + m_0^f \delta_{\alpha\beta})\psi^{\beta}_{f}(x), \tag{1.2}$$

$$S_{\mathrm{G}} = \frac{1}{4g_0^2} \int d^D x\, G^a_{\mu\nu}(x) G^a_{\mu\nu}(x). \tag{1.3}$$

Here, S_{mat} represents the fermion contribution, while S_{G} is the gluon kinetic term. Indices μ, ν denote the Lorentz ones, which run from 1 to 4 in the Euclidean space. The Dirac spinor $\psi^{\alpha}_{f}(x)$ with the flavor label $f = u, d, s, c, b, t$ has three color degrees of freedom labeled by $\alpha, \beta = 1, 2, 3$. $\psi^{\alpha}_{f}(x)$ belongs to the fundamental representation of the SU(3) Lie group. The bare mass is denoted by m_0^f for each flavor. The covariant derivative, which ensures the local gauge symmetry, is defined by

$$D_\mu = \partial_\mu + A^a_\mu(x) T^a, \tag{1.4}$$

with the bare gluon field $A^a_\mu(x)$ and Gell-Mann matrix T^a. The gluon field belongs to the $\mathfrak{su}(3)$ Lie algebra, and the indices a, b run 1 through 8. The field strength is made from the gauge field as follows:

$$G^a_{\mu\nu} = \partial_\mu A^a_\nu(x) - \partial_\nu A^a_\mu(x) + f^{abc} A^b_\mu(x) A^c_\nu(x). \tag{1.5}$$

Here, f^{abc} is the SU(3) structure constant and g_0 represents the bare gauge coupling. Since the last term in Eq. (1.5) represents the self-interaction among gluons with colors, it makes QCD as well as even the pure YM theory a very complicated theory in contrast to Abelian theories.

When the YM theory was first proposed, the renormalizability was not yet proved; hence, the theory looked somewhat skeptical. However, in 1971, 't Hooft proved the renormalizability [26], and moreover, Gross, Wilczek [27], and Politzer [28] showed the asymptotic freedom of the YM theory. By virtue of these establishments, QCD became the focus of attention. In the following, we overview the unique properties of QCD.

1.1.1 Asymptotic Freedom

In the quantum field theory, the coupling constant g depends on the renormalization scale μ as $g = g(\mu)$, where the scale μ is arbitrarily chosen because the theory does not contain it.[6] The self-interaction among gluons in the strong interaction plays a crucial role for the strong coupling constant to largely depend on the choice of the scale μ.

The μ-dependence of the coupling constant is given by

$$\mu \frac{\partial}{\partial \mu} g = \beta(g), \tag{1.6}$$

where $\beta(g)$ is referred to as the beta function. $\beta(g)$ can be calculated perturbatively when g is sufficiently small. In QCD, under the $\overline{\text{MS}}$ scheme, β is given by a series expansion with respect to g as

$$\beta(g) = -\beta_0 g^3 - \beta_1 g^5 + O(g^7), \tag{1.7}$$

$$\beta_0 = \frac{1}{(4\pi)^2}\left(11 - \frac{2}{3}N_f\right), \quad \beta_1 = \frac{1}{(4\pi)^4}\left(102 - \frac{38}{3}N_f\right), \tag{1.8}$$

where N_f denotes the number of flavors. As long as the condition $N_f \leq 16$ is satisfied, β is negative for small g. This implies that $g(\mu)$ decreases as μ becomes large, which is referred to as the asymptotic freedom [27, 28]. The asymptotic freedom is in sharp contrast to the case in quantum electrodynamics (QED), where the coupling constant $e(\mu)$ becomes large as μ increases. By using the beta function up to the leading order, $\beta(g) = -\beta_0 g^3$, we can solve Eq. (1.6) and explicitly obtain

$$\alpha_s(\mu) = \frac{g(\mu)^2}{4\pi} = \frac{1}{4\pi\beta_0 \ln(\mu^2/\Lambda_{\text{QCD}}^2)}. \tag{1.9}$$

---

[6] Thus, the physical quantities, such as the scattering cross section σ, do not depend on μ, which is summarized in the renormalization group equation as follows:

$$\mu \frac{d}{d\mu} \sigma(p, \mu, g(\mu)) = 0.$$

Here, p denotes the momentum in the external legs. Note that the quark mass dependence is suppressed. The renormalization equation provides the relation between the bare and renormalized quantities.

Here, the integral constant Λ_{QCD} is called the QCD scale parameter, which is determined experimentally. Equation (1.9) shows that $\alpha_s(\mu)$ diverges at $\mu \to \Lambda_{\text{QCD}}$.

1.1.2 Chiral Symmetry

Due to the asymptotic freedom of the coupling constant in QCD, when a physical process contains a large momentum transfer, the perturbative calculation is justified. On the other hand, the description of phenomena at low energy crucially requires a non-perturbative treatment. The spontaneous chiral symmetry breaking in QCD is one of such non-perturbative phenomena.

Let us now consider the left-handed and right-handed quarks defined as

$$\psi_{\text{L}} = \frac{1-\gamma_5}{2}\psi, \quad \psi_{\text{R}} = \frac{1+\gamma_5}{2}\psi, \tag{1.10}$$

where γ_5 represents the chirality operator. Here, we suppress the color and flavor indices. When masses of quarks are exactly zero, the QCD Lagrangian, Eq. (1.3), possesses the chiral symmetry under the global $\text{SU}(N_f)_{\text{L}} \otimes \text{SU}(N_f)_{\text{R}}$ transformation:

$$\psi_{\text{L}} \to e^{-\theta_{\text{L}}^a T^a}\psi_{\text{L}}, \quad \psi_{\text{R}} \to e^{-\theta_{\text{R}}^a T^a}\psi_{\text{R}}, \tag{1.11}$$

where θ_{L}^a and θ_{R}^a are parameters independent of space–time for the left-handed and right-handed quarks, respectively.

Since the masses of the u and d quarks are sufficiently smaller than Λ_{QCD}, we can regard that the chiral symmetry is approximately satisfied for the u and d quarks. Nevertheless, the proton or the neutron, which is composed of three valence u and d quarks, is much heavier than $3m_{u,d}$. This is attributed to the spontaneous chiral symmetry breaking.

Nambu and Jona-Lasinio proposed that the chiral symmetry is spontaneously broken in vacuum as [29, 30]

$$\text{SU}(N_f)_{\text{L}} \otimes \text{SU}(N_f)_{\text{R}} \to \text{SU}(N_f)_{\text{V}}, \tag{1.12}$$

by employing the idea of the BCS theory in the superconductivity [31]. This symmetry breaking results from the quark condensate $\langle\bar{\psi}(x)\psi(x)\rangle$, which is an analogy of the condensate of the Cooper pairs in superconductivity. When a symmetry is broken, there appears a massless mode, called Nambu–Goldstone mode. Nambu considered that the π meson corresponds to the resultant Nambu–Goldstone mode of the chiral symmetry breaking. In fact, the pion mass is much lighter than those of typical hadrons.

Since the nonzero quark condensate and the spontaneous symmetry breaking are definitely non-perturbative phenomena, perturbative calculations do not derive them

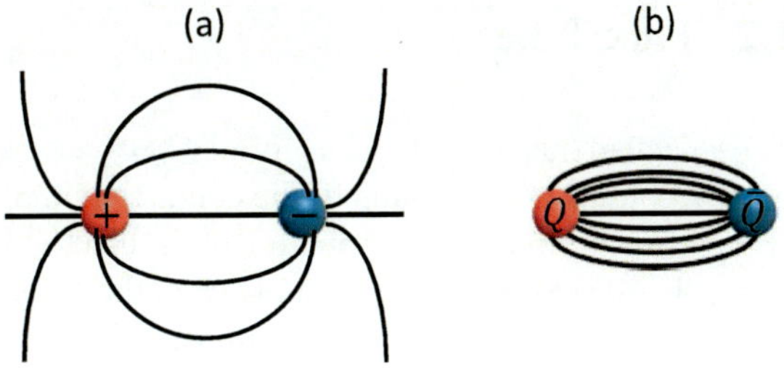

Fig. 1.1 Sketch of the distribution of (**a**) the electric field around two opposite charges and (**b**) the color electric field around Q and $\bar{Q}$. Each solid line corresponds to the line of force

from QCD. In order to reveal the non-perturbative aspects of QCD, the lattice QCD approach is useful. The lattice QCD calculation confirms the quark condensate with the use of the Banks–Casher relation [32–34].

In addition, we consider the thermal effect on the quark condensate. In the similar manner as the superconductivity is broken to the normal state at high temperature, the quark condensate is dissociated in hot medium, which leads to the restoration of the chiral symmetry. Therefore, QCD experiences the phase transition as the temperature becomes higher and the quark condensate plays a role of the order parameter [35].

1.1.3 Confinement

In addition to the dynamical chiral symmetry breaking in vacuum, the confinement of colors or especially quarks is another important property of QCD [36]. Quarks have fractional charges as mentioned above. If one can isolate a quark, a state with a fractional charge will be observed. However, no experiments with high accuracy have never observed any particles with fractional charges. This fact is referred to as the confinement of colors. This generally insists that the state which we can observe should be color singlet. Although there were a lot of attempts to mathematically prove the confinement on the basis of QCD, no works have not exactly achieved this challenging goal.

Nevertheless, these previous studies have provided phenomenological pictures on the confinement. An important description of the confinement is given by the flux tube, which is a one-dimensional tube structure connecting color charges. It is understood that the formation of the flux tube leads to the color confinement. As mentioned later, one of our main subjects discussed in this dissertation is the analysis on the local structure around the flux tube. Therefore, in the next section, we give a review on the flux tube in detail.

1.2 Flux Tube

The so-called flux tube, which is formed between a quark (Q) and an anti-quark ($\bar{Q}$), gives an intuitive description of the confinement. Shown in Fig. 1.1 is a sketch for a comparison between the distribution of the electric flux around two opposite charges in the classical Maxwell theory and that of the color electric flux around Q and $\bar{Q}$. In Fig. 1.1(a), when there is a pair of opposite electric charges, the electric field spreads all over the space. They interact with each other via the Coulomb force.

In contrast, the color electric flux induced by static $Q\bar{Q}$ charges is believed to be concentrated within a one-dimensional tube structure, the flux tube as shown in Fig. 1.1(b). The picture of the flux tube indicates a linearly rising potential proportional to the $Q\bar{Q}$ distance R for a large separation. In other words, the energy per unit length stored in the $Q\bar{Q}$ system is constant with respect to R, when separating $Q\bar{Q}$. Then, the infinitely large energy is required to isolate Q and $\bar{Q}$, which indicates the confinement.

When $Q\bar{Q}$ is sufficiently separated, the energy accumulated in the $Q\bar{Q}$ system suffices to create a new dynamical quark–anti-quark pair at some separation length. Thus, in fact, the system falls into a new color-singlet hadronic state to lower the energy in the system.

In order to understand the nature of the flux tube, there have been so far a lot of attempts based on effective models and numerical simulations from the first principle.

1.2.1 Dual Abelian-Higgs Model

Here, we introduce the dual Abelian-Higgs (DAH) model, which is one of the effective models to reveal the formation of the string structure between $Q\bar{Q}$. The DAH model is based on the dual superconductor picture proposed by Nambu [37], 't Hooft [38], and Mandelstam [39], which is an analogy with the ordinary superconductor.

Let us first consider the superconductivity in terms of the Ginzburg–Landau (GL) theory [40]. In the ordinary type-II superconductor, the magnetic monopole and anti-monopole are connected via a string because the magnetic flux is excluded from the non-perturbative vacuum with the condensation of the Cooper pairs. Abrikosov showed that the GL theory in fact possesses a vortex solution with a quantized magnetic flux in the case of type-II [41]. Inside the flux, the superconductor is partially broken, so that the normal state is realized.

The relativistic extension of the GL theory is the Abelian-Higgs (AH) model. Nielsen and Olesen showed that the AH model as well has a static vortex solution [42]. We notice here that while the Abrikosov vortex is a solution in the type-II superconductor, the AH model has a vortex solution whichever the type is. In order to capture the idea of the type in the superconductivity, we should consider that two vortices pull (push) each other in the case of type-I (type-II).

Now, we turn to the DAH model. The DAH model is related to the AH model under the dual transformation, which means the exchange between the electric field and the magnetic field as well as the exchange between the electric charge and the magnetic charge. In the viewpoint of the dual superconductivity, the electric flux is squeezed one-dimensionally in the vacuum with the magnetic monopole condensate. When we regard the electric charge as the color one, the confinement of quarks can be understood.

Although the dual superconductor picture mentioned above is intuitive, the justification of such a picture as an effective model for QCD is non-trivial. In fact, we should confront the following questions: (i) The dual transformation is not well defined in the non-Abelian theory. (ii) The color electric field can branch out. (iii) The existence of the magnetic monopole and its condensate is non-trivial.

In order to answer these questions, 't Hooft suggested that the Abelian projection or the Abelian gauge fixing to QCD leads us to the extraction of the Abelian sector with the monopole from the non-Abelian gauge theory [43]. In addition, under the Abelian dominance [44–46], which insists that the Abelian sector becomes crucial, it is confirmed that the monopole can condensate [47–49]. In particular, the maximal Abelian gauge among several (Abelian) gauge fixings is useful to realize the Abelian dominance [48]. Further details on the attempts to derive the DAH model from QCD are discussed in Refs. [50–54].

In this way, the formation of the string structure between $Q\bar{Q}$ is established.

1.2.2 Dynamics of Strings

Based on the existence of the string structure in $Q\bar{Q}$ systems via, for example, the DAH model, further studies on the dynamics of the string have been performed.

One of such attempts is the analysis with the Nambu–Goto (NG) string. The NG action or the slightly modified Polyakov action is proportional to the two-dimensional area swept by a string. By quantizing the theory, we obtain the oscillating open and closed strings, which correspond to infinite kinds of excited states. In particular, the first excited open string corresponds to the massless photon and the first excited closed string the graviton. Since the NG model is a bosonic one and requires the 26-dimensional space for the theoretical consistency, it is not necessarily compatible with the phenomenology of hadrons. Nevertheless, this direction is employed as a candidate of the theory of everything [55]; hence, there are a lot of active studies in this field.

As another approach, the world-sheet theory is the low energy effective string picture, proposed by Lüscher, Symanzik, and Weisz in 1980 [56]. The action of their model, called the LSW action, describes the transverse fluctuations of a string, which is induced by the spontaneous breaking of the translational invariance. Although there is a difference in the symmetries imposed on the LSW and NG actions, the LSW action corresponds to the derivative expansion of the NG action up to the next-to-next-to-leading order with the open–closed duality [57, 58]. The LSW model can

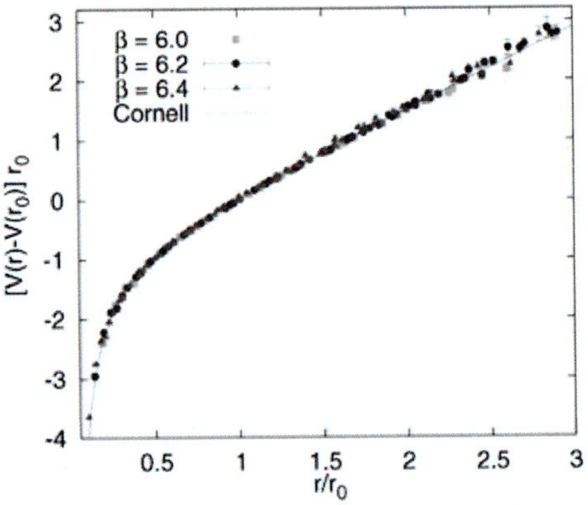

Fig. 1.2 $Q\bar{Q}$ potential calculated on the lattice based on the SU(3) YM gauge theory. The lattice data are well fitted by the Cornell potential. The figure is adapted from Ref. [64], Copyright 2001, with permission from Elsevier

explain the Coulombic term in the $Q\bar{Q}$ potential, which is often called the Lüscher term [59], as well as the liner term, and the logarithmical fattening of the flux tube with respect to the $Q\bar{Q}$ distance [60, 61].

It is true that the non-perturbative dynamics behind the confinement of quarks is difficult to understand, but to some extent the quantitative evaluation of the non-perturbative effect is feasible, focusing on, for example, the symmetry. In fact, the above picture is supported by the experimental observation such as the Regge trajectory [62].

1.2.3 Lattice Approach

For the fully quantitative and non-perturbative investigation from the first principle of QCD, we need the entirely new framework beyond the perturbation. To this effect, Wilson brought an innovative idea, that is, the lattice formalism of gauge theories [63]. The essential idea of the lattice formulation is to introduce the space–time lattice, where the lattice spacing plays a role of the ultraviolet cutoff in the regularization procedure. The introduction of the lattice structure makes uncountably infinite degrees of freedom in the calculation of the path integral countably infinite degrees of freedom. Then, by restricting the volume to be finite, we can numerically evaluate the physical quantity on the lattice from the first principle. In the actual calculation, we employ the Monte Carlo method with importance sampling because of the large number of degrees of freedom. Note that the volume has to be carefully chosen to be large enough not to severely affect the physics under consideration. Thanks to the recent development of the simulation algorism and the computer power, studies on the lattice are actively performed, including the investigation of the confinement problem. The details on the lattice formalism are discussed in Chap. 3.

Now, we turn to the studies on the flux tube in vacuum on the lattice. First of all, we should focus on the study on the $Q\bar{Q}$ potential. In Fig. 1.2, we show the $Q\bar{Q}$

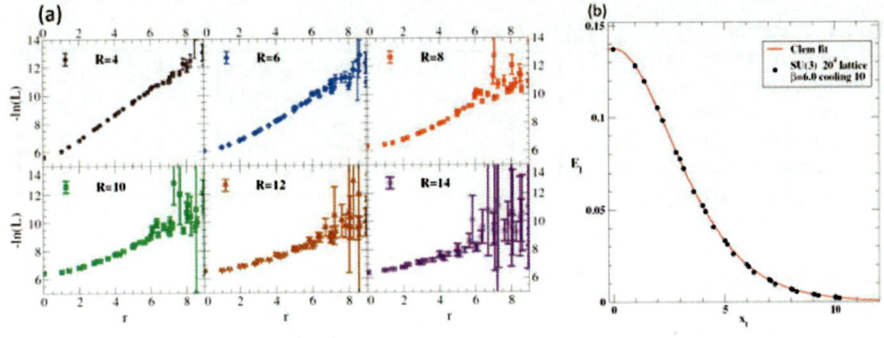

Fig. 1.3 Distribution of (**a**) the action density L and (**b**) the magnitude of the color electric field E_l around the flux tube in the SU(3) YM gauge theory. The horizontal axis in both figures represents the radial direction in the cylindrical coordinate. In the left figure, they also investigate the dependence of the $Q\bar{Q}$ distance R. In the right figure, the profile is fitted with the function obtained based on the dual superconductor picture with the assumption of infinitely long flux tube [69]. The left (right) figure is taken from Ref. [67] ([68]) with permissions, Copyright 2013 (2012) by the American Physical Society.

potential, which is calculated on the lattice based on the SU(3) YM gauge theory [64–66], where the standard Wilson loop is measured. The horizontal axis represents the static $Q\bar{Q}$ distance. In the same figure, the lattice result is well fitted with the Cornell parametrization as

$$V_{\text{Cornell}}(R) = -\mathcal{A}/R + \sigma R + \mathcal{B}, \tag{1.13}$$

where $\mathcal{A}$, σ and $\mathcal{B}$ are fit parameters. In particular, σ represents the string tension, which characterizes the linear confining force. At short distance, we find the Coulombic behavior. Since the perturbation is applicable at short distance due to the asymptotic freedom, this results mainly from the one-gluon exchange. On the other hand, one sees the linearly rising potential at large distance. This is attributed to the formation of the flux tube between Q and $\bar{Q}$. Therefore, such a direct observation of the linear-rising $Q\bar{Q}$ potential strongly suggests the existence of the flux tube.

After the establishment of the confinement potential on the lattice, analyses on the detailed structure of the flux tube have been carried out. We here introduce two examples of previous studies. Shown in Fig. 1.3 is the distribution of (a) the action density [67] and (b) the magnitude of the color electric field [68] around the flux tube in the SU(3) YM gauge theory. When we consider the $Q\bar{Q}$ system, it is useful to employ the cylindrical coordinates (r, θ, z), where z represents the direction connecting two charges and r, θ are radial and angular directions.[7] In Fig. 1.3, the horizontal axes labeled by r and x_t indicate the radial direction in the cylindrical coordinates. Thus, the figure shows the profile of those quantities as a function of the radial direction. These results are obtained from the measurement of the correlation

[7] See also Fig. 6.6 and refer to the detailed directions on the cylindrical coordinate in Chap. 6.

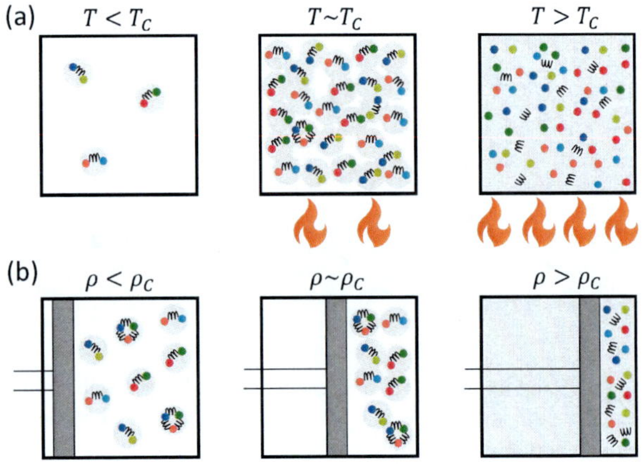

Fig. 1.4 Sketch of percolation of hadrons. (**a**) As the temperature T becomes high, hadrons are excited from the vacuum. Then, the hadrons have overlapped with one another, resulting in the percolation of quarks and gluons. (**b**) The percolation can also occur in the case that the density ρ is higher than the critical density ρ_c

function between the Wilson loop and the plaquette. In Fig. 1.3(a), Ref. [67] discusses the width of the flux tube. They suggest that the flux tube is logarithmically fattened with respect to the $Q\bar{Q}$ distance, which is consistent with the prediction in Refs. [60, 61]. In addition, as shown in Fig. 1.3(b), the profile is fitted with the function based on the dual superconductor picture. Reference [68] discusses the type of the vacuum based on the GL parameter under the assumption of the infinitely long flux tube [69]. We will revisit and thoroughly discuss these problems in this dissertation.

1.3 Phase Transition and Exploration of Deconfined Phase

As is briefly suggested above, QCD can experience the deconfinement phase transition as the temperature becomes higher. The high-temperature phase is called the deconfined phase, in which the plasma state composed of a lot of hot and colorful particles is realized. This state is referred to as the quark–gluon plasma (QGP). QGP is considered to have existed in the early universe for about 10^{-4} seconds after the Big Bang, where the temperature was more than 10^{12} K. A similar phase transition is also expected to occur in dense medium. There is a possibility for such a dense state to exist at the core of neutron stars. Historically, the existence of the deconfined phase was theoretically predicted in the superdense matter in 1975 by Collins and Perry [70].

The appearance of the deconfinement phase transition in the hot and dense medium can be understood in the following manner. At low temperature, QCD vacuum shows the confinement and the chiral symmetry breaking, so that the colorful states do not appear. On the other hand, at extremely high temperature, due to the asymptotic freedom, the picture of the approximately free gas composed of quarks and gluons is justified. Then, the quarks and gluons behave as fundamental degrees of freedom.

The phase transition is also understood intuitively from the viewpoint of the percolation of hadrons. Shown in Fig. 1.4 is a cartoon of the percolation of hadrons. In Fig. 1.4(a), we consider that the temperature of the system becomes higher. As the temperature becomes higher, only the lightest hadron, the pion, is in the beginning excited and the more massive hadrons are in order excited.[8] Since hadrons possess finite sizes, those thermally excited hadrons eventually overlap among one another. Then, constituent quarks and gluons percolate from their hadrons, so that the system undergoes the phase transition into the color conductor [71, 72]. The similar situation is expected to happen in the system at high baryon density. In Fig. 1.4(b), we consider that the baryon density becomes larger.

Now, we are tempted to ask questions about how the phase transition is, how high the transition temperature and the critical baryon density are, what the properties of the deconfined phase in QCD are, and so forth. These problems are of essentially non-perturbative, so that the lattice QCD calculation plays crucial roles as a theoretical attempt to calculate the observables.[9]

We here make remarks on the relativistic heavy ion collisions, which provide us a chance to experimentally explore such extreme conditions as QGP can be created. In order to produce the QGP and to investigate the physical properties of QGP, the experiments have been in progress at the Relativistic Heavy Ion Collider at Brookhaven National Laboratory since 2000 and at the Large Hadron Collider at European Organization for Nuclear Research, CERN, since 2009. So far, these experiments have presented us evidences of the existence of QGP. They suggest that QGP is a strongly coupled system and its behavior is almost described as a perfect fluid. In addition, unique properties of QGP have been found, such as the elliptic flow, jet quenching, heavy quarkonium suppression, and so on [73, 74].

From now on, we focus on the theoretical attempts to explore the deconfined system. One of the most important quantities in the exploration of the deconfined phase is the thermodynamic observable, such as the energy and the pressure in thermal equilibrium and homogeneous systems. The thermodynamic quantities provide us universal and essential information on the system. In Fig. 1.5, we show the lattice study on the thermodynamic quantities, (a) the entropy density and (b) the trace anomaly as a function of the temperature [75]. The numerical analyses are performed on the lattice based on the pure SU(3) YM gauge theory. Lattice studies on the thermodynamics have been actively performed as shown in Fig. 1.5, where each plot corresponds to different formalisms [75–80]. The results give us the information of

---

[8] In the pure YM gauge theory, there do not appear any hadrons and the only particle which can be excited in the confined phase is the massive glueball, whose mass is estimated to be $\gtrsim 1$ GeV. Since the glueball is more difficult to be excited than light hadrons in QCD, the transition temperature in the pure gauge theory is higher than that in QCD. Note that, in this sense, the transition temperature depends on the quark masses.

[9] Even at the high temperature slightly above the transition temperature, the perturbative calculation is not necessarily applicable. This is because the strong coupling is not small enough for the perturbation at $T = O(100)$ MeV. In addition, to make matters essentially difficult, the magnetic effect breaks down the perturbation at higher order due to the infrared divergence no matter how high the temperature is. In these senses, the non-perturbative formalism is crucial.

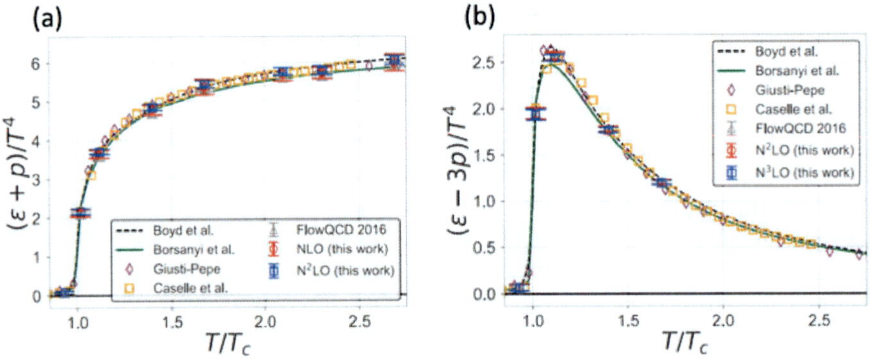

Fig. 1.5 Thermodynamic quantities as a function of the temperature. The numerical simulations are performed on the lattice based on the pure SU(3) YM gauge theory. (**a**) Equation of state of the entropy density. (**b**) Equation of state of the trace anomaly. In both figures, each plot corresponds to different methods to calculate the thermodynamic quantities. They all are almost consistent with one another. This figure is taken from Ref. [75] by The Author(s) licensed under CC BY 4.0 (http://creativecommons.org/licenses/by/4.0/)

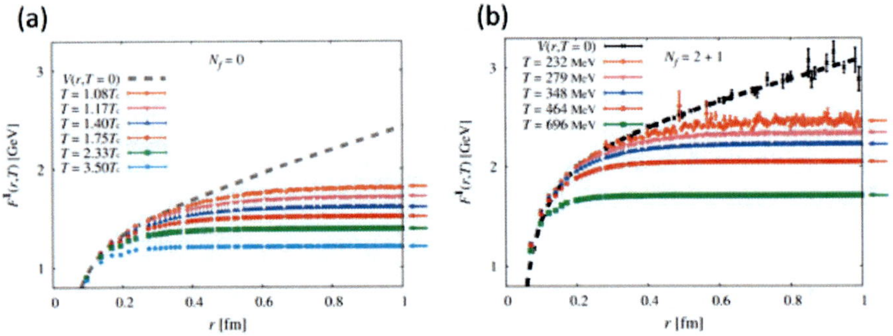

Fig. 1.6 Singlet free energy in the $Q\bar{Q}$ system at various temperatures above T_c. (**a**) In the quenched QCD calculation. (**b**) In the (2 + 1) flavor QCD calculation. Both figures are adapted from Ref. [81], Copyright 2012, with permission from Oxford University Press

the equation of state in the complicated non-perturbative system, which is of quite importance in the experimental search for QGP.

Next, we focus on the free energy in the $Q\bar{Q}$ system at finite temperature. This is a sharp comparison to the confinement potential observed in vacuum. Shown in Fig. 1.6 is the singlet free energy $F_1(r, T)$ in the static $Q\bar{Q}$ system with distance r at zero and various temperatures above T_c [81]. In Fig. 1.6(a), the numerical simulation is performed based on the pure SU(3) YM theory, while, in Fig. 1.6(b), the calculation is on the 2 + 1 flavor QCD. In the finite temperature lattice gauge theory, the singlet free energy is calculated via the correlation function between the Polyakov loop and its Hermitian conjugate as follows[10]:

[10] For the detail of the Polyakov loop, see Chap. 3.

$$e^{-F_1(r,T)/T} = \frac{1}{3}\mathrm{Tr}\langle \Omega(\mathbf{x}, T)\Omega^\dagger(\mathbf{0}, T)\rangle. \tag{1.14}$$

Here, $\Omega(\mathbf{x}, T)$ represents the Polyakov loop at site $\mathbf{x}$ and temperature T. The $Q\bar{Q}$ distance is given by $r = |\mathbf{x}|$. At short distance, the behaviors of the free energy are insensitive to the temperature. On the other hand, the asymptotic values at large distance correspond to twice the free energy of the single quark, $-T\ln\langle\mathrm{Tr}\Omega\rangle$, which depend on the temperature. In Fig. 1.6, the arrows denote them at each temperature. From the figure, we find that the confinement potential disappears at finite temperature above T_c, which suggests the deconfinement due to the screening by hot medium.

1.4 Purpose

QCD exhibits a large variety of unique phenomena as discussed above. The precise description of them is crucial to understand the strong interaction in nature. As overviewed in the previous sections, a lot of theoretical attempts to study phenomena in vacuum and hot medium have been performed on the basis of the lattice simulations and the effective models. However, our insights into those systems have been so far restricted within the following properties:

1. Global properties, such as the thermodynamics in Fig. 1.5 as well as the potential/free energy in Figs. 1.2 and 1.6.
2. Distributions whose physical meanings are not necessarily clear due to a lack of the gauge invariance as shown in Fig. 1.3(b).
3. Scalar and vector quantities, such as the action density and the magnitude of the color electric field.

The purpose of our study is to thoroughly investigate the vacuum and the deconfined phase of the SU(3) YM theory beyond the above limitations. To this effect, we focus on one of the most fundamental and important observables, the energy momentum tensor (EMT). We employ static test charges as probes to explore the complicated quantum systems such as the YM vacuum and study the response of the system against test charges through EMT.

EMT is a direct and fundamental physical quantity and has several remarkable advantages:

1. EMT is a special quantity among other observables in the sense that it relates the local properties and the global behaviors of the system in a gauge-invariant manner.
2. EMT has a definite physical meaning related to the energy density and the force acting on a surface. In addition, the absolute values of components of EMT are unambiguously meaningful thanks to the renormalization group invariance.

3. Since EMT is a second-rank tensor, it has many channels and hence provides more detailed information on the system, compared with the scalar and vector quantities.

For these reasons, EMT plays crucial roles as an important observable in various fields in physics including gravitational theory, hydrodynamics, elastic body, and so on. EMT can describe the local response under the external sources in a gauge-invariant manner, so that it provides us deep insights into the system in terms of the local interaction. In particular, EMT in non-Abelian gauge theory including QCD is even more important, where it is non-trivial to determine the direction of the line of force. It characterizes the local structure of the non-Abelian fields with gauge-invariant and non-perturbative information.

In our studies, we study the EMT distribution around static charges both in the lattice simulations based on the SU(3) YM gauge theory and in the effective model with the Abelian-Higgs (AH) model.

1.4.1 Lattice Study

We investigate the distribution of EMT in two specific systems in the lattice simulations. One target is the EMT distribution around a static single Q in the deconfined phase above T_c, and the other is that in the $Q\bar{Q}$ system in vacuum. Both simulations are performed based on the SU(3) YM gauge theory.

The practical use of EMT on the lattice has been in fact difficult because EMT is related to the translational invariance in space, which is explicitly broken on the lattice [82, 83]. Thanks to the recent development of the gradient-flow method [84–86] and its application to the EMT operator [87–89], we avoid this difficulty. The gradient flow is a kind of smearing, such as the ordinary diffusion equation. It has then turned out that the smearing property enables us to define the EMT operator, being independent of specific regularizations [87, 88]. This indicates that the EMT operator is defined even on the lattice, so that it gives the possibility of the numerical simulations on the lattice.

In fact, analyses of EMT defined via the gradient flow have been performed in various systems described by the strong interaction, such as the homogeneous thermal systems in the SU(3) YM theory [75, 76, 90–92] and in QCD [93, 94] and thermal system with pressure anisotropy [95]. In particular, the lattice study of thermodynamics in the SU(3) YM theory has been compared with those with other well-known methods, such as the integral methods and so on [75–80], so that the validity of EMT via the gradient flow has been confirmed. These results are summarized in Fig. 1.5. At the same time, the usefulness of the gradient flow has been revealed in that it requires smaller statistics for reasonable accuracies than other methods due to the noise reduction of the smearing of the fields.

They have investigated the global quantities, such as the pressure, energy density, entropy density, and specific heat. In this sense, our studies are the extension of these previous studies to the exploration in inhomogeneous systems with static probes. In the present studies, we provide novel aspects of the use of EMT via the gradient flow to investigate non-perturbative physics of the strong interaction.

In the study of the EMT distribution around a static Q in the deconfined phase of the SU(3) YM theory, we find that, although no significant difference can be seen between the EMT distribution in the radial and transverse directions in the spherical coordinate except for the sign, the temporal component, or the energy density, is substantially different from the spatial ones near the critical temperature T_c. This is a sharp contrast to the prediction of the classical Maxwell theory . The lattice data of the EMT distributions also indicate the thermal screening at long distance and the perturbative behavior at short distance.

We then proceed from the simpler system, a single Q, to the $Q\bar{Q}$ system although the present system is vacuum. We investigate the spatial distribution of EMT around $Q\bar{Q}$ pair in the SU(3) lattice gauge theory. The resultant EMT distribution with a decomposition into local principal axes shows the detailed structure of the flux tube along the longitudinal and transverse directions in a gauge invariant manner. We also find that the linear confining behavior of the $Q\bar{Q}$ potential at long distances is derived directly from the integral of the local stress tensor.

1.4.2 Effective Model

For the further investigation of the local structure around the flux tube, motivated by the numerical analysis of the $Q\bar{Q}$ system in the SU(3) YM gauge theory on the lattice, we study the EMT distribution around the flux tube based on the momentum conservation and the AH model. We note that we employ the magnetic monopoles and the magnetic vortex between them in the AH model instead of those in the DAH model since both models are essentially the same under the electromagnetic duality.

We first investigate constraints on the stress tensor distribution from the momentum conservation and show that the effect of boundaries plays a crucial role to describe the structure of the flux tube in the SU(3) YM theory obtained from the lattice simulation. We then study the distributions of EMT around the magnetic vortex with and without boundaries in the AH model and compare them with the distributions in the SU(3) YM theory based on the dual superconductor picture. It is shown that a wide parameter range of the AH model is excluded by a comparison with the lattice results in terms of EMT.

1.5 Organization

This dissertation is organized in the following manner. Chapters 2–4 are dedicated to the reviews of the theoretical tools in our studies.

In Chap. 2, the general properties of EMT are summarized. In particular, we visit the Maxwell theory and investigate the Maxwell stress tensor, which is an important example and comparison to understand our studies.

In Chap. 3, the review of the lattice gauge theory is given. In the final section, we introduce the gauge-invariant quantities on the lattice, which are measured in our lattice studies.

In Chap. 4, we recapitulate the gradient-flow method and its application to the definition of EMT, mainly focusing on the YM gauge theory. After a short review of the YM gauge theory and its renormalization in the ordinary D-dimensional space, we consider the basic idea of the gradient flow. The renormalizability of the YM gradient flow is also shown. Then, we proceed to the methodology to define the EMT via the gradient flow. At the end of this chapter, we briefly comment on the gradient flow in QCD.

After the above preparation, Chapters 5, 6 and 7 are the main body of this dissertation.

In Chap. 5, we study the EMT distribution around a static Q in the deconfined phase of the SU(3) YM theory on the lattice. The YM gradient flow plays crucial roles to define the EMT operator on the lattice and to explore its spatial structure.

In Chap. 6, we investigate the spatial EMT distribution around $Q\bar{Q}$ in the SU(3) YM gauge theory on the lattice. This chapter is roughly divided into two parts. In the first part, we consider the Wilson loop and determine an appropriate parameter set to enhance the signal of the observable. Next, with the use of such appropriate parameter sets, we investigate the EMT distribution in the $Q\bar{Q}$ system.

In Chap. 7, motivated by the study in Chap. 6, we study the EMT distribution around the flux tube in static $Q\bar{Q}$ systems based on the momentum conservation and the AH model. We first consider the momentum conservation, which is satisfied by EMT in general. We then show the numerical results of the EMT distribution around the magnetic vortex in the AH model with/without boundaries. The resultant EMT distribution around the magnetic vortex with finite length is compared with the lattice results obtained in Chap. 6.

Finally, Chap. 8 is devoted to the summary and outlook of our studies.

In the appendices, we provide auxiliary tools and notions for our studies.

In Appendix A, we explain the improved estimator, which is employed for the noise reduction. This technique plays useful roles in the lattice studies in Chapters 5 and 6.

In Appendix B, we discuss the procedure to make the tree-level improvement of the correlation between the EMT operator and the Polyakov loop on the lattice.

In Appendix C, the leading-order perturbative analysis of the correlation is presented using the high-temperature effective field theory, which is referred to as the electrostatic QCD (EQCD). Both Appendices B and C are auxiliary to Chap. 5.

In Appendix D, we introduce the smearing method to reduce the UV fluctuations of quantum fields, especially focusing on the APE smearing. This technique is mainly used in Chap. 6.

In Appendix E, we summarize the analytic properties of the infinitely long magnetic vortex in the AH model at the border of type-I and type-II, which is called the Bogomol'nyi bound. This appendix is mentioned in Chap. 7.

References

1. Rutherford E (1911) The scattering of alpha and beta particles by matter and the structure of the atom. Phil Mag Ser 6(21):669–688
2. Gegier H, Ernest M (1909) A diffuse reflection of the alpha particles. Proc R Soc Lond A 82:495–500
3. Rutherford E (2010) Collision of α particles with light atoms. IV. An anomalous effect in nitrogen. Phil Mag 90(sup1):31–37
4. Chadwick J (1932) Possible Existence of a Neutron. Nature 129:312
5. Yukawa H (1935) On the interaction of elementary particles I. Proc Phys Math Soc Jpn 17:4. [Prog. Theor. Phys. Suppl. **1**, 1]
6. Lattes CMG, Occhialini GPS, Powell CF (1947) Observations on the tracks of slow mesons in photographic emulsions. 1. Nature 160:453
7. Bloom ED et al (1969) High-energy inelastic e-p scattering at 6-degrees and 10-degrees. Phys Rev Lett 23:930
8. Breidenbach M et al (1969) Observed behavior of highly inelastic electron-proton scattering. Phys Rev Lett 23:935
9. Gell-Mann M (1964) A schematic model of baryons and mesons. Phys Lett 8:214
10. Zweig G, An SU(3) model for strong interaction symmetry and its breaking. Version 1. CERN-TH-401
11. Zyla PA et al (2020) [Particle Data Group], Review of Particle Physics. PTEP 2020(8):083C01
12. Glashow SL, Iliopoulos J, Maiani L (1970) Weak interactions with Lepton-Hadron symmetry. Phys Rev D 2:1285
13. Augustin JE et al (1974) [SLAC-SP-017 Collaboration], Discovery of a narrow resonance in e^+e^- annihilation. Phys Rev Lett 33:1406. [Adv. Exp. Phys. **5**, 141 (1976)]
14. Aubert JJ et al (1974) [E598 Collaboration], Experimental observation of a heavy particle J. Phys Rev Lett 33:1404
15. Kobayashi M, Maskawa T (1973) CP Violation in the Renormalizable Theory of Weak Interaction. Prog Theor Phys 49:652
16. Herb SW et al (1977) Observation of a dimuon resonance at 9.5-GeV in 400-GeV proton-nucleus collisions. Phys Rev Lett 39:252
17. Abe F et al (1995) [CDF Collaboration], Observation of top quark production in $\bar{p}p$ collisions. Phys Rev Lett 74:2626 [hep-ex/9503002]
18. Abachi S et al (1995) [D0 Collaboration], Search for high mass top quark production in $p\bar{p}$ collisions at $\sqrt{s} = 1.8$ TeV. Phys Rev Lett 74:2422. [hep-ex/9411001]
19. Han MY, Nambu Y (1965) Three triplet model with double SU(3) symmetry. Phys Rev 139:B1006
20. Swartz ML (1996) Reevaluation of the hadronic contribution to $\alpha(M_Z^2)$. Phys Rev D 53:5268 [hep-ph/9509248]
21. Yang CN, Mills RL (1954) Conservation of isotopic spin and isotopic gauge invariance. Phys Rev 96:191
22. Utiyama R (1956) Invariant theoretical interpretation of interaction. Phys Rev 101:1597–1607
23. Glashow SL (1961) Partial symmetries of weak interactions. Nucl Phys 22:579
24. Weinberg S (1967) A model of leptons. Phys Rev Lett 19:1264
25. Salam A (1968) Weak and electromagnetic interactions. Conf Proc C 680519:367
26. 't Hooft G, (1971) Renormalizable Lagrangians for massive Yang-Mills fields. Nucl Phys B 35:167
27. Gross DJ, Wilczek F (1973) Ultraviolet behavior of nonabelian Gauge theories. Phys Rev Lett 30:1343
28. Politzer HD (1973) Reliable perturbative results for strong interactions? Phys Rev Lett 30:1346
29. Nambu Y, Jona-Lasinio G (1961) Dynamical model of elementary particles based on an analogy with superconductivity. I. Phys Rev 122:345
30. Nambu Y, Jona-Lasinio G (1961) Dynamical model of elementary particles based on an analogy With superconductivity. II. Phys Rev 124:246

31. Bardeen J, Cooper LN, Schrieffer JR (1957) Theory of superconductivity. Phys Rev 108:1175–1204
32. Banks T, Casher A (1980) Chiral Symmetry breaking in confining theories. Nucl Phys B 169:103–125
33. Fukaya H et al (2010) [JLQCD] Determination of the chiral condensate from 2+1-flavor lattice QCD. Phys Rev Lett 104:122002. [erratum: Phys Rev Lett 105:159901] arXiv:0911.5555 [hep-lat]
34. Cossu G, Fukaya H, Hashimoto S, Kaneko T, Noaki JI (2016) Stochastic calculation of the Dirac spectrum on the lattice and a determination of chiral condensate in 2+1-flavor QCD. PTEP 2016(9):093B06. arXiv:1607.01099 [hep-lat]
35. Bazavov A, Bhattacharya T, Cheng M, DeTar C, Ding HT, Gottlieb S, Gupta R, Hegde P, Heller UM, Karsch F et al (2012) The chiral and deconfinement aspects of the QCD transition. Phys Rev D 85:054503 arXiv:1111.1710 [hep-lat]
36. Greensite J (2011) An introduction to the confinement problem. Lect Notes Phys 821:1–211
37. Nambu Y (1974) Strings, Monopoles and Gauge Fields. Phys. Rev. D 10:4262
38. 't Hooft G Gauge fields with unified weak, electromagnetic, and strong interactions. PRINT-75-0836 (UTRECHT)
39. Mandelstam S (1976) Vortices and quark confinement in nonabelian gauge theories. Phys Rept 23:245
40. Ginzburg VL, Landau LD (1950) On the Theory of superconductivity. Zh Eksp Teor Fiz 20:1064
41. Abrikosov AA (1957) The magnetic properties of superconducting alloys. J Phys Chem Solids 2:199
42. Nielsen HB, Olesen P (1973) Vortex line models for dual strings. Nucl Phys B 61:45
43. 't Hooft G, (1981) Topology of the gauge condition and new confinement phases in nonabelian gauge theories. Nucl Phys B 190:455
44. Ezawa ZF, Iwazaki A (1982) Abelian dominance and quark confinement in Yang-Mills theories. Phys Rev D 25:2681
45. Ezawa ZF, Iwazaki A (1982) Abelian dominance and quark confinement in Yang-Mills theories. 2. Oblique confinement and η' mass. Phys Rev D 26:631
46. Suzuki T, Yotsuyanagi I (1990) A possible evidence for Abelian dominance in quark confinement. Phys Rev D 42:4257–4260
47. Kronfeld AS, Schierholz G, Wiese UJ (1987) Topology and dynamics of the confinement mechanism. Nucl Phys B 293:461
48. Kronfeld AS, Laursen ML, Schierholz G, Wiese UJ (1987) Monopole condensation and color confinement. Phys Lett B 198:516
49. Maedan S, Suzuki T (1989) An infrared effective theory of quark confinement based on monopole condensation. Prog Theor Phys 81:229
50. Suzuki T (1988) A Ginzburg-Landau type theory of quark confinement. Prog Theor Phys 80:929
51. Ball JS, Caticha A (1988) Superconductivity: a testing ground for models of confinement. Phys Rev D 37:524
52. Maedan S, Matsubara Y, Suzuki T (1990) Abelian confinement mechanism and the QCD vacuum. Prog Theor Phys 84:130
53. Kodama H, Matsubara Y, Ohno S, Suzuki T (1997) Inter-meson potentials in dual Ginzburg-Landau theory. Prog Theor Phys 98:1345
54. Kondo KI, Kato S, Shibata A, Shinohara T (2015) Quark confinement: dual superconductor picture based on a non-Abelian Stokes theorem and reformulations of Yang-Mills theory. Phys Rept 579:1–226 arXiv:1409.1599 [hep-th]
55. Green MB, Schwarz JH (1984) Anomaly cancellation in supersymmetric D=10 gauge theory and superstring theory. Phys Lett B 149:117–122
56. Lüscher M, Symanzik K, Weisz P (1980) Anomalies of the free loop wave equation in the WKB approximation. Nucl Phys B 173:365
57. Lüscher M, Weisz P (2004) String excitation energies in SU(N) gauge theories beyond the free-string approximation. JHEP 07:014. arXiv:hep-th/0406205 [hep-th]

58. Aharony O, Karzbrun E (2009) On the effective action of confining strings. JHEP 06:012 arXiv:0903.1927 [hep-th]
59. Lüscher M (1981) Symmetry breaking aspects of the roughening transition in gauge theories. Nucl Phys B 180:317–329
60. Lüscher M, Munster G, Weisz P (1981) How thick are chromoelectric flux tubes? Nucl Phys B 180:1–12
61. Gliozzi F, Pepe M, Wiese UJ (2010) The width of the color flux tube at 2-loop order. JHEP 11:053 arXiv:1006.2252 [hep-lat]
62. Collins PDB (1977) An Introduction to Regge Theory and High-Energy Physics
63. Wilson KG (1974) Confinement of quarks. Phys Rev D 10:2445
64. Bali GS (2001) QCD forces and heavy quark bound states. Phys Rept 343:1–136. arXiv:hep-ph/0001312 [hep-ph]
65. Bali GS, Schilling K (1993) Running coupling and the Lambda parameter from SU(3) lattice simulations. Phys Rev D 47:661–672. arXiv:hep-lat/9208028 [hep-lat]
66. Bali GS, Schilling K, Wachter A (1997) Complete $O(v^2)$ corrections to the static interquark potential from SU(3) gauge theory. Phys Rev D 56:2566–2589. arXiv:hep-lat/9703019 [hep-lat]
67. Cardoso N, Cardoso M, Bicudo P (2013) Inside the SU(3) quark-antiquark QCD flux tube: screening versus quantum widening. Phys Rev D 88:054504 arXiv:1302.3633 [hep-lat]
68. Cea P, Cosmai L, Papa A (2012) Chromoelectric flux tubes and coherence length in QCD. Phys Rev D 86:054501 arXiv:1208.1362 [hep-lat]
69. Clem JR (1975) Simple model for the vortex core in a type II superconductor. J Low Temp Phys 18:427
70. Collins JC, Perry MJ (1975) Superdense matter: neutrons or asymptotically free quarks? Phys Rev Lett 34:1353
71. Baym G (1979) Confinement of quarks in nuclear matter. Phys A 96(1–2):131–135
72. Celik T, Karsch F, Satz H (1980) A percolation approach to strongly interacting matter. Phys Lett B 97:128–130
73. Andronic A, Arleo F, Arnaldi R, Beraudo A, Bruna E, Caffarri D, del Valle ZC, Contreras JG, Dahms T, Dainese A et al (2016) Heavy-flavour and quarkonium production in the LHC era: from proton-proton to heavy-ion collisions. Eur Phys J C 76(3):107 arXiv:1506.03981 [nucl-ex]
74. Aarts G, Aichelin J, Allton C, Arnaldi R, Bass SA, Bedda C, Brambilla N, Bratkovskaya E, Braun-Munzinger P, Bruno GE et al (2017) Heavy-flavor production and medium properties in high-energy nuclear collisions - What next?. Eur Phys J A 53(5):93. arXiv:1612.08032 [nucl-th]
75. Iritani T, Kitazawa M, Suzuki H, Takaura H (2019) Thermodynamics in quenched QCD: energy–momentum tensor with two-loop order coefficients in the gradient-flow formalism. PTEP 2019(2):023B02. arXiv:1812.06444 [hep-lat]
76. Kitazawa M, Iritani T, Asakawa M, Hatsuda T, Suzuki H (2016) Equation of state for SU(3) gauge theory via the energy-momentum tensor under gradient Flow. Phys Rev D 94(11):114512 arXiv:1610.07810 [hep-lat]
77. Boyd G, Engels J, Karsch F, Laermann E, Legeland C, Lutgemeier M, Petersson B (1996) Thermodynamics of SU(3) lattice gauge theory. Nucl Phys B 69:419-444. arXiv:hep-lat/9602007 [hep-lat]
78. Borsanyi S, Endrodi G, Fodor Z, Katz SD, Szabo KK (2012) Precision SU(3) lattice thermodynamics for a large temperature range. JHEP 07:056 [arXiv:1204.6184 [hep-lat]]
79. Giusti L, Pepe M (2017) Equation of state of the SU(3) Yang-Mills theory: A precise determination from a moving frame. Phys Lett B 769:385–390 [arXiv:1612.00265 [hep-lat]]
80. Caselle M, Nada A, Panero M (2018) QCD thermodynamics from lattice calculations with nonequilibrium methods: The SU(3) equation of state. Phys Rev D 98(5):054513 arXiv:1801.03110 [hep-lat]
81. Maezawa Y, Umeda T, Aoki S, Ejiri S, Hatsuda T, Kanaya K, Ohno H (2012) Application of fixed scale approach to static quark free energies in quenched and 2+1 flavor lattice QCD with improved Wilson quark action. Prog Theor Phys 128:955–970 arXiv:1112.2756 [hep-lat]

82. Caracciolo S, Curci G, Menotti P, Pelissetto A (1988) The energy momentum tensor on the lattice: the scalar case. Nucl Phys B 309:612
83. Caracciolo S, Curci G, Menotti P, Pelissetto A (1990) The energy momentum tensor for lattice gauge theories. Ann Phys 197:119
84. Narayanan R, Neuberger H (2006) Infinite N phase transitions in continuum Wilson loop operators. JHEP 0603:064 [hep-th/0601210]
85. Lüscher M (2010) Properties and uses of the Wilson flow in lattice QCD. JHEP 1008:071 arXiv:1006.4518 [hep-lat]
86. Lüscher M, Weisz P (2011) Perturbative analysis of the gradient flow in non-abelian gauge theories. JHEP 1102:051 arXiv:1101.0963 [hep-th]
87. Suzuki H (2013) Energy-momentum tensor from the Yang-Mills gradient flow. PTEP 2013(8):083B03. [Erratum: PTEP 2015(7):079201 (2015)] arXiv:1304.0533 [hep-lat]
88. Makino H, Suzuki H (2015) Lattice energy–momentum tensor from the Yang–Mills gradient flow—inclusion of fermion fields. PTEP 2014:063B02. [erratum: PTEP 2015:079202 (2015)] arXiv:1403.4772 [hep-lat]
89. Harlander RV, Kluth Y, Lange F (2018) The two-loop energy–momentum tensor within the gradient-flow formalism. Eur Phys J C 78(11):944 [erratum: Eur Phys J C 79(10):858 (2019)] arXiv:1808.09837 [hep-lat]
90. Asakawa M et al (2014) [FlowQCD], Thermodynamics of SU(3) gauge theory from gradient flow on the lattice. Phys Rev D 90(1):011501 [erratum: Phys Rev D 92(5):059902 (2015)] arXiv:1312.7492 [hep-lat]
91. Kitazawa M, Iritani T, Asakawa M, Hatsuda T (2017) Correlations of the energy-momentum tensor via gradient flow in SU(3) Yang-Mills theory at finite temperature. Phys Rev D 96(11):111502 arXiv:1708.01415 [hep-lat]
92. Shirogane M, Ejiri S, Iwami R, Kanaya K, Kitazawa M, Suzuki H, Taniguchi Y, Umeda T (2021) Latent heat and pressure gap at the first order deconfining phase transition of SU(3) Yang-Mills theory using the small flow-time expansion method. PTEP 2021(1):013B08. arXiv:2011.10292 [hep-lat]
93. Taniguchi Y, Ejiri S, Iwami R, Kanaya K, Kitazawa M, Suzuki H, Umeda T, Wakabayashi N (2017) Exploring N_f = 2+1 QCD thermodynamics from the gradient flow. Phys Rev D 96(1):014509. [erratum: Phys Rev D 99(5):059904 (2019)] arXiv:1609.01417 [hep-lat]
94. Taniguchi Y et al (2020) [WHOT-QCD], N_f = 2+1 QCD thermodynamics with gradient flow using two-loop matching coefficients. Phys Rev D 102(1):014510 arXiv:2005.00251 [hep-lat]
95. Kitazawa M, Mogliacci S, Kolbé I, Horowitz WA (2019) Anisotropic pressure induced by finite-size effects in SU(3) Yang-Mills theory. Phys Rev D 99(9):094507 arXiv:1904.00241 [hep-lat]

Chapter 2
Energy Momentum Tensor

Symmetry is a key notion to grasp the nature in a unified manner. It restricts the laws of the nature and somewhat plays a role to make the nature what it is in the actual world. Noether's theorem [1], in particular, involves in continuous symmetries, which states the mathematically rigorous relation between the symmetry and the conserved quantity. According to the theorem, the invariance of an action under a global or local transformation results in the existence of a conserved quantity and vice versa. Moreover, in the quantum theory, the Ward–Takahashi (WT) identity [2, 3] generally ensures that the existence of symmetry such as the gauge symmetry leads to some relation of Green functions, vertex functions, and so on. In this sense, the WT identity is a quantum counterpart to Noether's theorem in classical theory.

Above all, in the field theory, the energy momentum tensor (EMT) $\mathcal{T}_{\mu\nu}(x)$ $(\mu, \nu = 0, 1, 2, 3)$, which is the conserved current based on the space–time symmetry, plays crucial roles in a broad range of fields in physics, such as gravitational theory, hydrodynamics, and elastic body. Since EMT is a gauge-invariant quantity, it enables us to directly explore the global properties as well as the local properties of the system even in the highly complicated system, such as the quantum Yang–Mills (YM) vacuum.

In this chapter, we recapitulate the general properties of EMT mainly based on Ref. [4]. After the derivation of EMT according to Noether's theorem in Sect. 2.1.1, we will discuss the physical meaning of the components of EMT in Sect. 2.1.2. In Sect. 2.1.3, we especially focus on the EMT in the gauge theory. Sect. 2.2 is dedicated to EMT in the Maxwell theory as a specific example, which will be helpful for the later chapters where our studies are discussed. We notice that, throughout this chapter, we consider the $(3 + 1)$-dimensional Minkowski space with the metric $g^{\mu\nu} = g_{\mu\nu} = \mathrm{diag}(1, -1, -1, -1)$.

R. Yanagihara, *Distribution of Energy Momentum Tensor around Static Charges in Lattice Simulations and an Effective Model*, Springer Theses,
https://doi.org/10.1007/978-981-16-6234-8_2

2.1 Derivation and Properties of Energy Momentum Tensor

2.1.1 *Derivation and Properties*

According to Noether's theorem, when an action $S[\varphi]$ is invariant under the translation $x^\mu \to x^\mu + \delta x^\mu$, we find out the conserved current $\mathcal{T}^{\mu\nu}_{\text{can}}$, which satisfies the conservation law,

$$\partial_\mu \mathcal{T}^{\mu\nu}_{\text{can}} = 0. \tag{2.1}$$

Here, we generally denote the field of interest as φ and the action is given by $S[\varphi] = \int d^4x\, \mathcal{L}[\varphi, \partial_\mu \varphi]$. $\mathcal{T}^{\mu\nu}_{\text{can}}$ is known as the canonical EMT and is explicitly written down as

$$\mathcal{T}^{\mu\nu}_{\text{can}} = \frac{\delta \mathcal{L}}{\delta(\partial_\mu \varphi)} \partial^\nu \varphi - g^{\mu\nu} \mathcal{L}, \tag{2.2}$$

which satisfies the condition,

$$\frac{\delta S[\varphi]}{\delta \varphi} \delta\varphi + \partial_\mu j^\mu = 0 \quad \text{with} \quad j^\mu = -\mathcal{T}^{\mu\nu}_{\text{can}} \delta x_\nu. \tag{2.3}$$

The vanishing divergence Eq. (2.1) indicates that there exists a conserved four-momentum, which is given by the integration over the hypersurface $x^0 = \text{const.}$,

$$P^\mu = \int_S dS_k\, \mathcal{T}^{\mu k}, \tag{2.4}$$

where S_k represents the hypersurface $x^0 = \text{const.}$ with $k = 1, 2, 3$. In this definition, we assume that the integral on the surface vanishes due to the absence of any charges. The 0th component P^0 is referred to as the canonical Hamiltonian or the total energy. One can directly confirm $P^0 = H$ since $\mathcal{T}^{00}_{\text{can}} = (\partial \mathcal{L}/\partial \dot{\varphi})\dot{\varphi} - \mathcal{L} = \mathcal{H}$ is just the Legendre transformation of $\mathcal{L}$. The spatial component P^i $(i = 1, 2, 3)$ is regarded as the momentum in three-dimensional space. Soon after, we will consider the meaning of each component in detail.

In the definition of EMT, there is an ambiguity. In fact, we always have a freedom to add the divergence of an antisymmetric tensor, so-called superpotential term, to the canonical EMT,

$$\mathcal{T}^{\mu\nu}_{\text{imp}} = \mathcal{T}^{\mu\nu}_{\text{can}} + \partial_\rho \psi^{\mu\rho\nu}, \tag{2.5}$$

where the superpotential term satisfies $\psi^{\mu\rho\nu} = -\psi^{\rho\mu\nu}$. Since the anti-symmetry of $\psi^{\mu\rho\nu}$ guarantees the conservation of $\partial_\rho \psi^{\mu\rho\nu}$, $\mathcal{T}^{\mu\nu}_{\text{imp}}$ is also conserved; hence, the four-momentum remains unchanged. This means that the four-momentum is a unique quantity. As a matter of fact, the Lorentz invariance requires the symmetry of the EMT.[1]

In order to give a unique definition for EMT, we impose the specific condition that the density of the four angular momentum, or so-called Lorentz current, is represented by the ordinary formula with the momentum density:

$$M^{\mu\rho\sigma} = x^\rho \mathcal{T}^{\mu\sigma} - x^\sigma \mathcal{T}^{\mu\rho}. \tag{2.6}$$

In this case, the total angular momentum is denoted by the four-momentum as well. We note that, in contrast, the canonical angular momentum $M^{\mu\rho\sigma}_{\text{can}}$ is by no means related to the canonical EMT $\mathcal{T}^{\mu\nu}_{\text{can}}$. It is easy to find out the condition on EMT to satisfy Eq. (2.6). Indeed, the conservation law for the angular momentum, $\partial_\mu M^{\mu\rho\sigma} = 0$, requires EMT to be symmetric as $\mathcal{T}^{\rho\sigma} = \mathcal{T}^{\sigma\rho}$. Here we use the conservation law for EMT and $\partial x^\mu / \partial x^\nu = \delta^\mu_\nu$. Thus, by choosing the superpotential term $\psi^{\mu\rho\nu}$ appropriately, we can render the EMT symmetric.

We comment on another derivation of the symmetric EMT. The trick to obtain the symmetric tensor directly is the differential of the action with respect to the metric,

$$\frac{1}{2}\sqrt{-g}\mathcal{T}^{\mu\nu} = \frac{\delta S}{\delta g_{\mu\nu}} = \frac{\partial \mathcal{L}}{\partial g_{\mu\nu}} - \partial^\rho \frac{\partial \mathcal{L}}{\partial(\partial^\rho g_{\mu\nu})}, \tag{2.7}$$

where g represents the determinant of the metric. In the following discussion, whatever procedure we use to define the EMT, we should employ the physical EMT that is symmetric in its indices and, in particular, gauge invariant in the gauge theory.

2.1.2 *Physical Meanings*

In the previous section, we introduced the formal derivation and the properties of EMT. Here, we consider the physical meanings of each component in EMT.

Among the components of EMT, $\mathcal{T}_{00}$ is interpreted as the energy density and $\mathcal{T}_{0i}$ $(i = 1, 2, 3)$ is the momentum and energy flow density.[2] From the conservation law, focusing on the 0th component, we obtain the following relation:

$$\frac{\partial \mathcal{T}^{00}}{\partial t} + \frac{\partial \mathcal{T}^{0i}}{\partial x^i} = 0, \tag{2.8}$$

with $x^0 = t$, or equivalently,

---

[1] In the Euclidean space, the rotational invariance ensures the symmetry of the EMT with respect to its indices.

[2] In the Euclidean space, the energy density is represented as $-\mathcal{T}_{44}$ after the Wick rotation.

$$0 = \frac{\partial}{\partial t}\int_V \mathcal{T}^{00} dV + \int_V \frac{\partial \mathcal{T}^{0i}}{\partial x^i} dV = \frac{\partial}{\partial t}\int_V \mathcal{T}^{00} dV + \oint_{S=\partial V} \mathcal{T}^{0i} dS_i, \tag{2.9}$$

after the integration over a volume V. Here, $S = \partial V$ denotes the surface of V with the outgoing surface vector. This equation suggests that the variation in the amount of the energy in the volume V is equal to the energy flow going out through the surface S.

Next, the spatial part $\mathcal{T}_{ij}$ $(i, j = 1, 2, 3)$ is related to the stress tensor σ_{ij} with [4]

$$\sigma_{ij} = -\mathcal{T}_{ij}, \tag{2.10}$$

in which σ_{ii} represents the pressure while σ_{ij} $(i \neq j)$ the shear stress. The force per unit area $\mathcal{F}_i$ acting on a surface with the normal vector n_i is given by the stress tensor as

$$\mathcal{F}_i = \sigma_{ij} n^j = -\mathcal{T}_{ij} n^j. \tag{2.11}$$

The local principal axes $n_i^{(k)}$ and the corresponding eigenvalues λ_k of the local stress tensor are obtained by solving the eigenvalue problem,

$$\mathcal{T}_i^j n_j^{(k)} = \lambda_k n_i^{(k)} \quad (k = 1, 2, 3). \tag{2.12}$$

The strength of the force per unit area along $n_i^{(k)}$ is given by the absolute values of the eigenvalue λ_k. Neighboring volume elements separated by a surface with the normal vector $n_i^{(k)}$ pull (push) each other for $\lambda_k < 0$ $(\lambda_k > 0)$ across the boundary. Note that the three principal axes $n_i^{(k)}$ are orthogonal with one another because σ_{ij} is a symmetric tensor.

In the same manner as the relation between the energy density and the momentum flow, Eq. (2.8) or (2.9), is derived, one can confirm the equilibrium of the force:

$$\frac{\partial \mathcal{T}^{i0}}{\partial t} + \frac{\partial \mathcal{T}^{ij}}{\partial x^j} = 0, \tag{2.13}$$

or

$$0 = \frac{\partial}{\partial t}\int_V \mathcal{T}^{i0} dV + \int_V \frac{\partial \mathcal{T}^{ij}}{\partial x^j} dV = \frac{\partial}{\partial t}\int_V \mathcal{T}^{i0} dV + \int_{S=\partial V} \mathcal{T}^{ij} dS_j. \tag{2.14}$$

2.1.3 Example: Energy Momentum Tensor in Pure Gauge Theory

Now let us apply the above discussion to the pure YM gauge theory, whose action is given by[3]

$$S_{\mathrm{G}} = \frac{1}{4g_0^2}\int d^4x\, G^a_{\mu\nu}(x)G^{\mu\nu\, a}(x). \tag{2.15}$$

Then, we derive the canonical EMT according to Eq. (2.2),

$$\mathcal{T}^{\mu\nu}_{\mathrm{can}} = \frac{1}{g_0^2}\left[G^{\mu\rho\, a}\partial^\nu A^a_\rho - \frac{1}{4}g^{\mu\nu}G^a_{\rho\sigma}G^{\rho\sigma\, a}\right]. \tag{2.16}$$

As Noether's theorem states, $\mathcal{T}^{\mu\nu}_{\mathrm{can}}$ accurately satisfies the conservation law Eq. (2.1) as it should do. Here, we discuss the features which Eq. (2.16) should retain, in addition to the conservation:

1. Gauge invariance:
   Equation (2.16) does not enjoy the gauge invariance under the infinitesimal transformation $\delta A^a_\mu(x) = \partial_\mu w^a(x) + f^{abc}A^b_\mu(x)w^c(x)$ with some gauge function $w^a(x)$. Thus, Eq. (2.16) is unsuitable for the observable.
2. Symmetry with respect to indices and Lorentz invariance:
   As commented in the previous part, the canonical EMT is not symmetric, so that the canonical angular momentum is not represented as Eq. (2.6). In other words, Eq. (2.16) does not have the Lorentz symmetry.

In order to avoid these unacceptable features, we construct the improved EMT which is symmetric in its indices as well as gauge invariant.[4] By choosing the superpotential term $\psi^{\mu\rho\nu} = G^{\mu\rho\, a}A^{\nu\, a}$, we obtain the improved EMT,

$$\mathcal{T}^{\mu\nu}_{\mathrm{imp}} = G^{\mu\rho\, a}G^{\rho\, a}_{\nu} - \frac{1}{4}g^{\mu\nu}G^{\rho\sigma\, a}G^a_{\rho\sigma}. \tag{2.17}$$

From Eq. (2.17) without further transformations, we can confirm the tracelessness $\mathcal{T}^\mu_\mu = 0$, which is a remarkable consequence from the fact that the present theory holds the conformal symmetry in addition to the scale symmetry.[5] In fact, we can construct the generators of both the scale transformation and the conformal transformation.

---

[3] In this dissertation, we assign $F_{\mu\nu}(x)$ ($G^a_{\mu\nu}(x)$) to the field strength in the Abelian (non-Abelian) theory. In the Abelian case, of course, the color index a does not appear.

[4] Although the improved EMT which respects the symmetry in its indices does not necessarily ensure the gauge invariance in a direct manner, this choice of the superpotential term in the gauge theory plays a role of the gauge covariantization as well as the symmetrization.

[5] In general, one considers 'new' improved EMT to convert EMT into the traceless one.

When we proceed to the quantum field theory, those symmetries are explicitly broken due to the quantum effect [5, 6]. This is referred to as the trace anomaly, which is represented by

$$\partial_\mu \Delta^\mu = \mathcal{T}^\mu_\mu = \frac{\beta}{2g} G^a_{\mu\nu} G^{\mu\nu a}, \tag{2.18}$$

in the chiral limit. Here, β represents the beta function and $\Delta^\mu \equiv x_\nu \mathcal{T}^{\nu\mu}$ is the dilatation current. The trace anomaly indicates that the translational symmetry and the scale invariance are not compatible with each other in the quantum theory.

2.2 Energy Momentum Tensor in Maxwell Theory

In order to grasp the idea of EMT in more detail, we review the Maxwell theory and the Maxwell stress. As indicated above, the Maxwell stress is naturally derived from the conservation of energy and momentum. We remark that the Maxwell theory is not only a foundation of physics since undergraduate, but cardinal in that, from the viewpoint of the conformal theory, the Maxwell theory in vacuum is the first case, which is confirmed to have the conformal symmetry [7, 8].

2.2.1 Maxwell Equations and Classical System

In the classical mechanism, we start with the Newton equation and the set of the Maxwell equations. When the position of a point charge e is denoted by $\mathbf{r}(t)$ at time t, the charge density $\rho(\mathbf{x}, t)$ and the electric current $\mathbf{j}(\mathbf{x}, t)$ are given by

$$\rho(\mathbf{x}, t) = e\delta^{(3)}(\mathbf{x} - \mathbf{r}(t)), \tag{2.19}$$

$$\mathbf{j}(\mathbf{x}, t) = e\dot{\mathbf{r}}(t)\delta^{(3)}(\mathbf{x} - \mathbf{r}(t)). \tag{2.20}$$

They accurately satisfy the conservation law of the electric charge $\partial_t \rho(\mathbf{x}, t) + \nabla \cdot \mathbf{j}(\mathbf{x}, t) = 0$. The force acting on the charge is the Lorentz force including the self-interaction,

$$\mathbf{F} = e\mathbf{E}(\mathbf{r}(t), t) + e\dot{\mathbf{r}}(t) \times \mathbf{B}(\mathbf{r}(t), t), \tag{2.21}$$

where $\mathbf{E}(\mathbf{r}(t), t)$ and $\mathbf{B}(\mathbf{r}(t), t)$ represent the electric and magnetic field, respectively. We notice that the time evolution of the electric and magnetic field is induced by both explicit t-dependence and implicit t-dependence via $\mathbf{r}(t)$. The former contribution represents the time evolution of distribution of the field itself. The latter results from the motion of the particle, that is, the change of $\mathbf{r}(t)$, even if the distribution of the field itself does not change.

Now, we consider the system composed of N electric particles, whose masses and electric charges are m_i and e_i $(i = 1, 2, \ldots, N)$, respectively. The motion of a particle is determined according to the Newton equation. Under the existence of the electric and magnetic field, the motions of the particles are denoted by

$$m_i \frac{d^2\mathbf{r}_i(t)}{dt^2} = e_i \mathbf{E}(\mathbf{r}_i(t), t) + e_i \dot{\mathbf{r}}_i(t) \times \mathbf{B}(\mathbf{r}_i(t), t). \tag{2.22}$$

In addition, since the existence of the charges induces the electric and magnetic field, their dependence on t and x follows the four sets of equations, so-called Maxwell equations:

$$\nabla \cdot \mathbf{D}(\mathbf{x}, t) = \sum_{i=1}^{N} e_i \delta^{(3)}(\mathbf{x} - \mathbf{r}_i(t)), \tag{2.23}$$

$$\nabla \cdot \mathbf{B}(\mathbf{x}, t) = 0, \tag{2.24}$$

$$\nabla \times \mathbf{E}(\mathbf{x}, t) + \frac{\partial \mathbf{B}(\mathbf{x}, t)}{\partial t} = 0, \tag{2.25}$$

$$\nabla \times \mathbf{H}(\mathbf{x}, t) - \frac{\partial \mathbf{D}(\mathbf{x}, t)}{\partial t} = \sum_{i=1}^{N} e_i \delta^{(3)}(\mathbf{x} - \mathbf{r}_i(t)), \tag{2.26}$$

where we employ the electric flux density $\mathbf{D}(\mathbf{x}, t)$ and the magnetic field $\mathbf{H}(\mathbf{x}, t)$.

Therefore, what we have to do is to solve the simultaneous equations composed of N Newton equations and the Maxwell equations with the unknown function $\mathbf{E}(\mathbf{x}, t)$ and $\mathbf{B}(\mathbf{x}, t)$. This set of equations completely describes the classical system under an initial condition, so that we can, in principle, grasp all the information on the system. However, due to the extremely complicated structure of the nonlinear simultaneous differential equations, even today it is too difficult a question to obtain the general solution of them. In most cases of interest, we handle the problems under the approximations, such as the small variance of the electric and magnetic field, and/or the small reaction that the electric charge and the current receive after the electromagnetic radiation.

2.2.2 Energy Conservation

In spite of the difficulty in solving the set of equations, we can acquire knowledge related to the conservation of energy. By performing the energy integral of the equation of motion (EoM), we have

$$\sum_i m_i \mathbf{v}_i(t) \cdot \frac{d\mathbf{v}_i(t)}{dt} = \sum_i \int_V d^3x\, e_i \delta^{(3)}(\mathbf{x} - \mathbf{r}_i(t))[\mathbf{v}_i(t) \cdot \mathbf{E}(\mathbf{x}, t) + \mathbf{v}_i(t) \cdot \{\mathbf{v}_i(t) \times \mathbf{B}(\mathbf{x}, t)\}]$$
$$\sum_i \frac{d}{dt}\left(\frac{1}{2} m_i \mathbf{v}_i^2(t)\right) = \sum_i \int_V d^3x\, e_i \delta^{(3)}(\mathbf{x} - \mathbf{r}_i(t)) \mathbf{v}_i(t) \cdot \mathbf{E}(\mathbf{x}, t), \tag{2.27}$$

where we use the fact that the magnetic field is always perpendicular to the direction of motion. For a further transformation, we apply the Ampère-Maxwell law, Eq. (2.26):

$$\sum_i \frac{d}{dt}\left(\frac{1}{2} m_i \mathbf{v}_i^2(t)\right) = \int_V d^3x \left(\nabla \times \mathbf{H}(\mathbf{x}, t) - \frac{\partial \mathbf{D}(\mathbf{x}, t)}{\partial t}\right) \cdot \mathbf{E}(\mathbf{x}, t). \tag{2.28}$$

Now, we consider the following quantity with the Faraday law, Eq. (2.25):

$$\begin{aligned} \frac{1}{2}\frac{\partial}{\partial t}(\mathbf{E} \cdot \mathbf{D} + \mathbf{B} \cdot \mathbf{H}) &= \mathbf{E} \cdot \frac{\partial \mathbf{D}}{\partial t} + \mathbf{H} \cdot \frac{\partial \mathbf{B}}{\partial t} \\ &= \mathbf{E} \cdot \frac{\partial \mathbf{D}}{\partial t} - \mathbf{H} \cdot (\nabla \times \mathbf{E}). \end{aligned} \tag{2.29}$$

This quantity allows us to write down Eq. (2.28) as

$$\frac{d}{dt}\left[\sum_i \frac{1}{2} m_i \mathbf{v}_i^2(t) + \frac{1}{2}\int_V d^3x\, (\mathbf{E} \cdot \mathbf{D} + \mathbf{B} \cdot \mathbf{H})\right] = -\int_V d^3x\, \nabla \cdot (\mathbf{E} \times \mathbf{H}). \tag{2.30}$$

In order to make the physical meaning more explicit, we have

$$-\frac{d}{dt}\left(\sum_i \frac{1}{2} m_i \mathbf{v}_i^2(t) + W(t)\right) = \oint_{S=\partial V} \mathbf{S}(\mathbf{x}, t) \cdot \hat{n}(\mathbf{x})\, dS, \tag{2.31}$$

where we define

$$W(t) = \int_V d^3x\, w(\mathbf{x}, t), \tag{2.32}$$
$$w(\mathbf{x}, t) = \frac{1}{2}(\mathbf{E}(\mathbf{x}, t) \cdot \mathbf{D}(\mathbf{x}, t) + \mathbf{B}(\mathbf{x}, t) \cdot \mathbf{H}(\mathbf{x}, t)), \tag{2.33}$$
$$\mathbf{S}(\mathbf{x}, t) = \mathbf{E}(\mathbf{x}, t) \times \mathbf{H}(\mathbf{x}, t), \tag{2.34}$$

where $\mathbf{S}(\mathbf{x}, t)$ is referred to as the Poynting vector. In the left-hand side of Eq. (2.31), the first term represents the kinetic energy of all the charges, while the second indicates the whole energy of the electromagnetic field. In the viewpoint of the local interaction, the existence of the charge induces the distortion of the vacuum, so that the energy of the electromagnetic field is accumulated. Thus, Eq. (2.31) means that the variation of the total energy carried by the point charges as well as the electromagnetic field is equal to the energy flow $\oint_{S=\partial V} \mathbf{S}(\mathbf{x}, t) \cdot \hat{n}(\mathbf{x})\, dS$ going out through $S = \partial V$. Notice that $\mathbf{S} \neq 0$ does not necessarily means the energy flow. In fact, static electric and magnetic fields lead to $\mathbf{S}(\mathbf{x}, t) \neq 0$ with no energy flow.

2.2.3 *Momentum Conservation*

In addition to the energy conservation, we see that the total momentum in the system is conserved as well. When we remove the electric charge and current from EoM with the use of the Maxwell equation, we have

$$\frac{d}{dt}\mathbf{G}_m(t) = \int_V d^3x \left[\mathbf{E}(\mathbf{x}, t)\nabla \cdot \mathbf{D}(\mathbf{x}, t) + \left(\nabla \times \mathbf{H}(\mathbf{x}, t) - \frac{\partial \mathbf{D}(\mathbf{x}, t)}{\partial t} \right) \times \mathbf{B}(\mathbf{x}, t) \right], \tag{2.35}$$

where we define the total momentum of all the charges as

$$\mathbf{G}_m(t) := \sum_i m_i \mathbf{v}_i(t). \tag{2.36}$$

Then, using the relation,

$$\frac{\partial}{\partial t}(\mathbf{D} \times \mathbf{B}) = \frac{\partial \mathbf{D}}{\partial t} \times \mathbf{B} + \mathbf{D} \times \frac{\partial \mathbf{B}}{\partial t} = \frac{\partial \mathbf{D}}{\partial t} \times \mathbf{B} - \mathbf{D} \times (\nabla \times \mathbf{E}), \tag{2.37}$$

the above equation is further transformed as

$$\begin{aligned} &\frac{d}{dt}\left(\mathbf{G}_m(t) + \frac{1}{c^2} \int_V d^3x\, \mathbf{S}(\mathbf{x}, t) \right) \\ &= \int_V d^3x\, [\mathbf{E}(\mathbf{x}, t)\nabla \cdot \mathbf{D}(\mathbf{x}, t) - \mathbf{D}(\mathbf{x}, t) \times (\nabla \times \mathbf{E}(\mathbf{x}, t)) - \mathbf{B}(\mathbf{x}, t) \times (\nabla \times \mathbf{H}(\mathbf{x}, t))]. \end{aligned} \tag{2.38}$$

In order to clarify the meaning, we define the following tensors with respect to the electric and magnetic field:

$$\mathcal{T}_{ij}(\mathbf{x}, t) = \mathcal{T}^e_{ij}(\mathbf{x}, t) + \mathcal{T}^m_{ij}(\mathbf{x}, t), \tag{2.39}$$

$$\mathcal{T}^e_{ij}(\mathbf{x}, t) = \varepsilon_0 \left(E_i(\mathbf{x}, t)E_j(\mathbf{x}, t) - \frac{1}{2}\delta_{ij}\mathbf{E}^2(\mathbf{x}, t) \right), \tag{2.40}$$

$$\mathcal{T}^m_{ij}(\mathbf{x}, t) = \frac{1}{\mu_0} \left(B_i(\mathbf{x}, t)B_j(\mathbf{x}, t) - \frac{1}{2}\delta_{ij}\mathbf{B}^2(\mathbf{x}, t) \right). \tag{2.41}$$

Equation (2.39) is referred to as the Maxwell stress, which is also expressed in the same manner as Eq. (2.17) except for the suppression of the color indices. The divergence of each tensor gives

$$(\nabla T^e)_k = \varepsilon_0[E_k \nabla \cdot \mathbf{E} - (\mathbf{E} \times (\nabla \times \mathbf{E}))_k], \tag{2.42}$$

$$(\nabla T^m)_k = \frac{1}{\mu_0}[B_k \nabla \cdot \mathbf{B} - (\mathbf{B} \times (\nabla \times \mathbf{B}))_k] = -\frac{1}{\mu_0}(\mathbf{B} \times (\nabla \times \mathbf{B}))_k, \tag{2.43}$$

so that the relation Eq. (2.38) reads

$$\frac{d}{dt}[\mathbf{G}_m(t) + \mathbf{G}_f(t)] = \int_V d^3x \, \nabla \mathcal{T}(\mathbf{x}, t) = \oint_{S=\partial V} dS \, \mathcal{T}(\mathbf{x}, t) \cdot \hat{n}(\mathbf{x}), \tag{2.44}$$

with the total momentum of the electromagnetic field

$$\mathbf{G}_f(t) = \frac{1}{c^2} \int_V d^3x \, \mathbf{S}(\mathbf{x}, t). \tag{2.45}$$

From Eq. (2.44), we find that the increase in the total momentum $\mathbf{G}_m + \mathbf{G}_f$ is denoted by the force $\oint_S dS \mathcal{T} \cdot \hat{n}$ acting from outside of the system V through the closed surface S. In other words, $\mathcal{T} \cdot \hat{n}$ represents the electromagnetic stress acting on a unit surface of S in vacuum, which is nothing but the distortion of vacuum in terms of the local interaction. In particular, if the whole system is closed, the total momentum is conserved.

2.2.4 *Maxwell Stress and Coulomb Force*

As a specific case, we consider the system with static charges, where there is only a static electric field. In general, the force per ΔS of the closed surface S is given by

$$F_1 = (\mathcal{T}^e_{11} n_1 + \mathcal{T}^e_{12} n_2 + \mathcal{T}^e_{13} n_3)\Delta S, \tag{2.46}$$

$$F_2 = (\mathcal{T}^e_{21} n_1 + \mathcal{T}^e_{22} n_2 + \mathcal{T}^e_{23} n_3)\Delta S, \tag{2.47}$$

$$F_3 = (\mathcal{T}^e_{31} n_1 + \mathcal{T}^e_{32} n_2 + \mathcal{T}^e_{33} n_3)\Delta S. \tag{2.48}$$

Here, $\hat{n} = (n_1, n_2, n_3)$ is an outgoing normal vector perpendicular to ΔS.

We consider the stress acting on the tube of electric force in Fig. 2.1. $\Delta S^{(1)}$ is an infinitely small surface perpendicular to the electric field $\mathbf{E}$, so that $\mathbf{E} \parallel \hat{n}^{(1)}$, or $\mathbf{E} = |\mathbf{E}|\hat{n}^{(1)}$. Then, since the Maxwell stress $\mathcal{T}^e_{ij}$ reads

$$\mathcal{T}^e_{11} = \frac{\varepsilon_0}{2}(E_1^2 - E_2^2 - E_3^2) = \frac{\varepsilon_0 E^2}{2}((n_1^{(1)})^2 - (n_2^{(1)})^2 - (n_3^{(1)})^2), \tag{2.49}$$

$$\mathcal{T}^e_{12} = \varepsilon_0 E_1 E_2 = \varepsilon_0 E^2 n_1^{(1)} n_2^{(1)}, \tag{2.50}$$

$$\mathcal{T}^e_{13} = \varepsilon_0 E_1 E_3 = \varepsilon_0 E^2 n_1^{(1)} n_2^{(1)}, \tag{2.51}$$

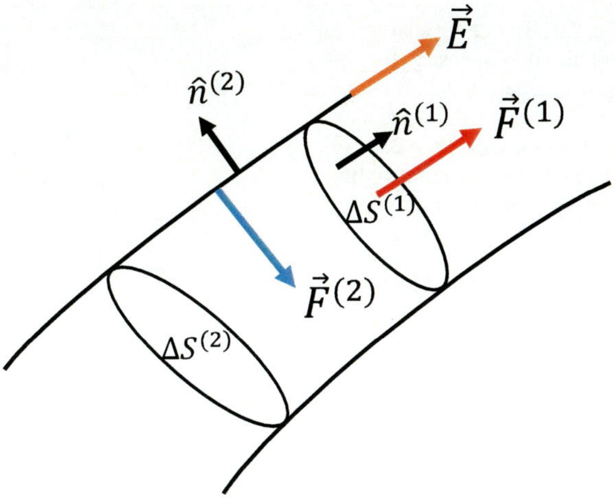

Fig. 2.1 Maxwell stress acts on the tube of electric force. $\Delta S^{(1,2)}$ represents infinitely small surface perpendicular to the electric field **E**. The normal vectors $\hat{n}^{(1,2)}$ are chosen to be $\mathbf{E} \parallel \hat{n}^{(1)}$ and $\mathbf{E} \perp \hat{n}^{(2)}$

we obtain the force along x direction on the fragment $\Delta S^{(1)}$

$$\begin{aligned} F_1^{(1)} &= \frac{\varepsilon_0 E^2}{2}[((n_1^{(1)})^2 - (n_2^{(1)})^2 - (n_3^{(1)})^2)n_1^{(1)} + 2((n_2^{(1)})^2 + (n_2^{(1)})^2)(n_1^{(1)})]\Delta S^{(1)} \\ &= \frac{\varepsilon_0 E^2}{2} n_1^{(1)} \Delta S^{(1)}. \end{aligned} \tag{2.52}$$

Thus, totally the force $\mathbf{F}^{(1)}$ is expressed as

$$\mathbf{F}^{(1)} = \frac{\varepsilon_0 E^2}{2} \hat{n}^{(1)} \Delta S^{(1)}, \quad |\mathbf{F}^{(1)}| > 0, \tag{2.53}$$

which indicates that there acts the tension in the outward direction on the equipotential surface. In the physical viewpoint, the neighboring volume elements pull each other, so that the length of the electric tube be as short as possible.

Next, we investigate the force acting on the side of the tube of electric force. $\Delta S^{(2)}$ and $\hat{n}^{(2)}$ are the infinitely small element on the side of the surface and the outgoing normal vector perpendicular to $S^{(2)}$, such that $\mathbf{E} \perp \hat{n}^{(2)}$, or $\mathbf{E} \cdot \hat{n}^{(2)} = 0$. In the same manner as discussed above, we obtain the force on $\Delta S^{(2)}$ as

$$\mathbf{F}^{(2)} = -\frac{\varepsilon_0}{2} E^2 \hat{n}^{(2)} \Delta S^{(2)}, \quad |\mathbf{F}^{(2)}| < 0, \tag{2.54}$$

which means the pushing force or pressure between the neighboring tubes.

Finally, we show the equivalence between the Maxwell stress and the Coulomb force in the static case. As shown in Fig. 2.2, we assume that there are two static charges e_1 and e_2 on the z-axis, whose separation distance is $2a$. We obtain the force acting on e_1 by summing up the stress over any closed surface which includes e_1. In particular, we have an option to choose the infinitely large surface perpendicular to

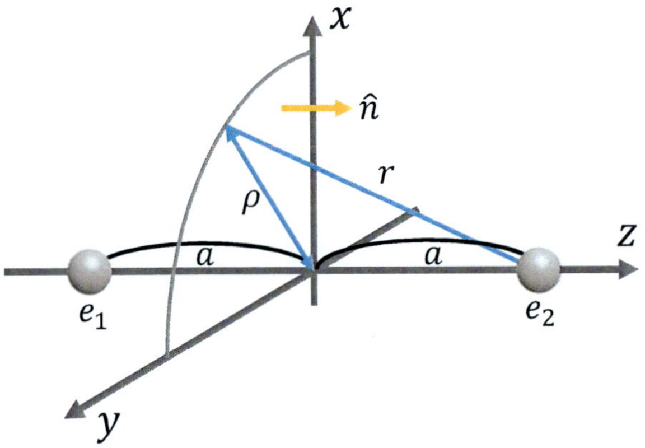

Fig. 2.2 Two static charges and the coordinates. Two static charges, e_1 and e_2, exist on the z-axis, which are separated at a distance $2a$. The origin is at the middle of them. ρ (r) represents lengths between the origin (each charge) and a point on $z = 0$, or the xy plane

z direction, that is, xy plane,

$$F_z = \int_{xy\text{-plane}} \mathcal{T} \cdot \hat{n} dS. \tag{2.55}$$

Since the electric field is given by

$$E_z(x) = \frac{1}{4\pi\varepsilon_0}\left(\frac{e_1}{r^2}\frac{a}{r} - \frac{e_2}{r^2}\frac{a}{r}\right) = \frac{e_1 - e_2}{4\pi\varepsilon_0}\frac{a}{r^3}, \tag{2.56}$$

$$E_\rho(x) = \frac{1}{4\pi\varepsilon_0}\left(\frac{e_1}{r^2}\frac{\rho}{r} + \frac{e_2}{r^2}\frac{\rho}{r}\right) = \frac{e_1 + e_2}{4\pi\varepsilon_0}\frac{\rho}{r^3}, \tag{2.57}$$

where ρ and r are indicated in Fig. 2.2 and $\hat{n}$ is parallel to z direction. Then, we have the explicit form of the stress

$$\mathcal{T}_{zz}(x) = \frac{\varepsilon_0}{2}(E_z^2 - E_\rho^2) = \frac{1}{32\pi^2\varepsilon_0(a^2+\rho^2)^3}[(e_1 - e_2)^2 a^2 - (e_1 + e_2)^2\rho^2], \tag{2.58}$$

which accurately leads to the ordinary Coulomb force

$$F_z = \int_S dS\,\mathcal{T}_{zz} = \int_0^\infty \mathcal{T}_{zz} \cdot 2\pi\rho d\rho = -\frac{1}{4\pi\varepsilon_0}\frac{e_1 e_2}{(2a)^2}. \tag{2.59}$$

2.2.5 *Stress Distribution on Source Plane*

In order to grasp the notion of the stress, especially the Maxwell stress, more intuitively, we show the distribution of the Maxwell stress on the plane where two opposite charges exist. This is a crucial comparison to our result in the SU(3) YM gauge

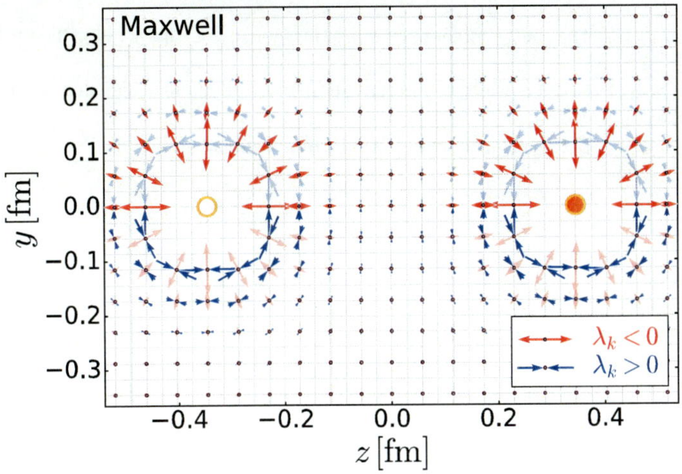

Fig. 2.3 Distribution of the Maxwell stress on the source plane. The red (blue) arrows correspond to the eigenvectors with negative (positive) eigenvalues, which are highlighted in the upper (lower) half panel

theory, which we discuss in the later chapter. Shown in Fig. 2.3 is the two eigenvectors in Eq. (2.12) along the principal axes of the local Maxwell stress. The other eigenvector is perpendicular to the yz plane. For the later comparison, the source is separated at distance $16\bar{a}$ with $\bar{a} = 0.029$ fm. The red (blue) arrow is highlighted on the upper (lower) plane in the figure for just an aesthetic reason. The negative (positive) eigenvalue is represented by the red outward (blue inward) arrow whose length is proportional to $\sqrt{|\lambda_k|}$:

$$\leftarrow\!\circ\!\rightarrow \,:\, \lambda_k < 0, \qquad \rightarrow\!\circ\!\leftarrow \,:\, \lambda_k > 0. \tag{2.60}$$

Here, $\sqrt{|\lambda_k|}$ corresponds to the electric field up to the constant factor, rather than the stress itself. Neighboring volume elements are pulling (pushing) with each other along the direction of red (blue) arrows. From Fig. 2.3, we find that the propagation of force spreads all over the space.

References

1. Noether E (1918) Invariant variation problems. Gott Nachr 1918:235–257. arXiv:physics/0503066 [physics]
2. Ward JC (1950) An identity in quantum electrodynamics. Phys Rev 78:182
3. Takahashi Y (1957) On the generalized Ward identity. Nuovo Cim 6:371
4. Landau LD, Lifshitz EM (1980) The classical theory of fields, 4 edn. Butterworth-Heinemann
5. Collins JC, Duncan A, Joglekar SD (1977) Trace and dilatation anomalies in gauge theories. Phys Rev D 16:438–449

6. Nielsen NK (1977) The energy momentum tensor in a nonabelian quark gluon theory. Nucl Phys B 120:212–220
7. Cunningham E (1910) The principle of relativity in electrodynamics and an extension thereof. Proc Lond Math Soc s2-8(1):77–98
8. Bateman H (1910) The transformation of the electrodynamical equations. Proc Lond Math Soc s2-8(1):223–264

Chapter 3
Lattice Field Theory

The fundamental theory of the elementary particle physics is written in terms of the quantum field theory (QFT). In the standard model, which describes the fundamental interactions in nature, the electromagnetic, weak, and strong interaction, quantum chromodynamics (QCD) is a gauge theory for the strong interaction on the dynamics of quarks and gluons. Since QCD contains self-interactions of gauge particles in contrast to quantum electrodynamics (QED), there exhibit diverse phenomena with the unique mechanisms such as the confinement, the spontaneous chiral symmetry breaking, and the asymptotic freedom as introduced in Chap. 1. Although QCD is established as the theory of strong interaction while exposed to plenty of strict experimental investigations, the large strong coupling in addition to the nonlinearity of the theory makes it difficult that, from the theoretical aspects, one explores the physics at low energy region with the use of the perturbative method. In fact, when the momentum transfer Q^2 is lower than $O(1)\,\mathrm{GeV}^2$, the strong coupling is large enough that the perturbative expansion diverges; hence, the perturbative analysis is unreliable. In addition, there are possibilities of missing any effects which are not detected in the perturbative analyses.

Beyond the framework of the perturbative expansion which is successful in QED as well as asymptotically high-energy QCD, the systematic method embracing the non-perturbative effect is the lattice formulation originally proposed in Ref. [1]. This technique is based on the lattice regularization, where the space and time are discretized into hypercubic so that the lattice spacing plays a role of regulator. Quarks and gluons are subject to each site and each link joining them on the lattice, respectively. Then, the Feynman path integral to calculate the partition functions and correlation functions reduces to the large dimensional ordinary integrals, whose calculation is feasible by Monte Carlo simulations on large-scale parallel computer systems. After the first proposal of the lattice technique by Wilson [1], this direction was adapted to study the lattice gauge theories [2, 3]. The simulation involved in quarks also started in Ref. [4]. Until now, with remarkable improvements of com-

R. Yanagihara, *Distribution of Energy Momentum Tensor around Static Charges in Lattice Simulations and an Effective Model*, Springer Theses,
https://doi.org/10.1007/978-981-16-6234-8_3

puting power as well as that of numerical algorithms, non-perturbative studies at the high precision level have achieved great successes for the better understanding of QCD.

The purpose of this chapter is to introduce the basic idea of the lattice formalism based on Refs. [5–7]. First, we review the quantization via the Feynman path integral in Sect. 3.1. In Sect. 3.2, we discuss the lattice regularization of both free boson systems and free fermion systems. We encounter the lattice gauge theory with the gauge principle in Sect. 3.3. In Sect. 3.4, we define gauge-invariant quantities on the lattice which are keys for our studies later.

3.1 Feynman Path Integral

In classical mechanics, the motion of a particle is described as a solution of a classical equation of motion, which is given by the saddle point of the action under the potential $V(q)$,

$$S = \int_{t_\mathrm{i}}^{t_\mathrm{f}} dt\, L(q, \dot{q}) = \int_{t_\mathrm{i}}^{t_\mathrm{f}} dt \left(\frac{1}{2} \left(\frac{dq}{dt} \right)^2 - V(q) \right). \tag{3.1}$$

Here, $L(q, \dot{q})$ represents the Lagrangian of this system as functions of the position q and the velocity $\dot{q} = dq/dt$. In this case, the classical particle propagates from $q = q_\mathrm{i}$ at the initial time $t = t_\mathrm{i}$ to the destination $q = q_\mathrm{f}$ at $t = t_\mathrm{f}$ along the definite classical path, subjected to the force $-dV/dq$.

On the other hand, in quantum mechanics, all possible paths connecting between $q = q_\mathrm{i}$ and $q = q_\mathrm{f}$ contribute to the particle propagation as quantum fluctuations. The information on the quantum system is denoted by the probability amplitude $\langle q_\mathrm{f}, t_\mathrm{f} | q_\mathrm{i}, t_\mathrm{i} \rangle = \langle q_\mathrm{f} | e^{-i\hat{H}(t_\mathrm{f}-t_\mathrm{i})} | q_\mathrm{i} \rangle$, where $\hat{H}$ denotes the Hamiltonian operator. In the Euclidean space, by replacing t with $-i\tau$, the amplitude is given by $\langle q_\mathrm{f}, \tau_\mathrm{f} | q_\mathrm{i}, \tau_\mathrm{i} \rangle = \langle q_\mathrm{f} | e^{-\hat{H}(\tau_\mathrm{f}-\tau_\mathrm{i})} | q_\mathrm{i} \rangle$. The main idea of the Feynman path integral is that, on the basis of the superposition principle, we perform the integration over all possible paths with the Boltzmann weight $e^{-S_\mathrm{E}[q]}$ as follows [8]:

$$\langle q_\mathrm{f}, \tau_\mathrm{f} | q_\mathrm{i}, \tau_\mathrm{i} \rangle = \int_{q_\mathrm{i}}^{q_\mathrm{f}} \mathcal{D}q\, e^{-S_\mathrm{E}[q]}, \tag{3.2}$$

where the Euclidean action is written by

$$S_\mathrm{E}[q] = \int_{t_\mathrm{i}}^{t_\mathrm{f}} dt\, L_\mathrm{E}(q, \dot{q}) = \int_{t_\mathrm{i}}^{t_\mathrm{f}} dt \left(\frac{1}{2} \left(\frac{dq}{d\tau} \right)^2 + V(q) \right). \tag{3.3}$$

Note that the classical case is reproduced in the $\hbar \to 0$ limit. In the following, since we concentrate on the quantities in the Euclidean space, we abbreviate the subscript 'E' unless otherwise mentioned.

We now proceed to QFT. We can regard QFT as the quantum mechanics with the infinitely large number of degrees of freedom, $\boldsymbol{\phi}(\mathbf{x}, \tau) = e^{\hat{H}\tau}\boldsymbol{\phi}(\mathbf{x}, 0)e^{-\hat{H}\tau}$ with the continuous parameter $\mathbf{x} \in \mathbb{R}^3$. All the information on the quantum system is inscribed into the time-ordered vacuum expectation value of $\boldsymbol{\phi}(x)$. In particular, the partition function is obtained in the same manner as above:

$$\mathcal{Z}(T) = \int \mathcal{D}\phi \, e^{-S[\phi]}. \tag{3.4}$$

The periodic boundary condition $\phi(0) = \phi(\beta)$ with $t \in [0, \beta]$ prescribes the temperature $T = 1/\beta$. The l-point Green function with the vacuum $|\Omega\rangle$,

$$G^{(l)}(x_1, \ldots, x_l) = \langle \Omega | \mathcal{T}[\boldsymbol{\phi}(x_1) \cdots \boldsymbol{\phi}(x_l)] | \Omega \rangle, \tag{3.5}$$

is also written down in the path integral representation as

$$G^{(l)}(x_1, \ldots, x_l) = \frac{\int \mathcal{D}\phi \, \phi(x_1) \cdots \phi(x_l) e^{-S[\phi]}}{\int \mathcal{D}\phi \, e^{-S[\phi]}}. \tag{3.6}$$

We notice that there emerges a time-ordered product of the operators in Eq. (3.5) while there does not in Eq. (3.6) due to the reduction to the c-number.

We finally comment on the path integral for the fermion field in the case we handle the matter fields with spin half. Basically, the expression of the path integral for the fermion field is similar to the bosonic one except for the fact that the fermion fields anti-commute with each other according to the Grassmann algebra.

3.2 Lattice Regularization

The system discussed in terms of QFT has uncountable infinite degrees of freedom denoted by $\mathbf{x}$, while the path integral formalism of the Green function is well defined for the system with countable degrees of freedom. Thus, the path integral in QFT discussed above is defined only formally. In order to make the path integral well defined, we consider the discretization of space and time, hence the space–time lattice. The introduction of the lattice corresponds to the regularization, where the inverse of the lattice spacing plays a role of regulator for the UV cutoff in the momentum space, just like the well-known Pauli–Villars mass and the dimensional regularization parameter ϵ. We need to remove the lattice structure ultimately so that we obtain the renormalized quantity observed in the laboratory. We refer to this process in the renormalization program as the continuum limit. It is true that the bare quantities strongly depend on the cutoff after the renormalization process, but the observables do not. In this section, after we introduce the procedure of the lattice discretization, we apply the lattice formalism to the boson and fermion field.

3.2.1 Simple Example

As a simple example, we feature a square-integrable function $f(x)$. We define the Fourier transformation of $f(x)$ and its inverse as

$$f(x) = \int_{-\infty}^{\infty} \frac{dk}{2\pi} \tilde{f}(k) e^{ikx}, \quad \tilde{f}(x) = \int_{-\infty}^{\infty} dx\, f(x) e^{-ikx}. \tag{3.7}$$

When we let x take only the value $x = na$ with the integer n and the lattice spacing a, $f(x)$ is decomposed into

$$f(na) = \int_{-\pi/a}^{\pi/a} \frac{dk}{2\pi} \tilde{f}_a(k) e^{ikna}, \tag{3.8}$$

where the periodicity is satisfied as $\tilde{f}_a(-\pi/a) = \tilde{f}_a(\pi/a)$. This indicates that the momentum integration is restricted within the range $[-\pi/a, \pi/a]$. The inverse Fourier transformation is also written as

$$\tilde{f}_a(k) = a \sum_{n=-\infty}^{\infty} f(na) e^{-ikna}. \tag{3.9}$$

From the above simple example, the introduction of the lattice spacing gives the cutoff of the momentum integration. In particular, we mention the delta function $\delta(x)$. By taking $f(na) = 1/(2\pi)$, the Fourier decomposition is expressed by

$$\delta(k) = \frac{a}{2\pi} \sum_{n} e^{-ikna}, \tag{3.10}$$

under the periodic boundary condition. In the configuration space, the delta function reduces to the Kronecker delta

$$\delta_{nm} = a \int_{-\pi/a}^{\pi/a} \frac{dk}{2\pi} e^{i(n-m)a}. \tag{3.11}$$

3.2.2 Free Boson Field

We consider the free boson system represented by the action,

$$S[\phi] = \frac{1}{2}\int d^4x\,(-\partial_\mu\partial_\mu + m_0^2)\phi(x), \tag{3.12}$$

where $\phi(x)$ denotes the real-valued scalar field in the Euclidean space and m_0 is the bare mass. μ runs 1 through 4. Then, the partition function and the l-point Green function are represented in the path integral form as follows:

$$\mathcal{Z} = \int \mathcal{D}\phi\, e^{-S[\phi]}, \tag{3.13}$$

$$G^{(l)}(x_1, \ldots, x_l) = \frac{1}{\mathcal{Z}}\int \mathcal{D}\phi\, \phi(x_1)\cdots\phi(x_l)e^{-S[\phi]}. \tag{3.14}$$

So far, they are not yet well defined but are formally written down. Thus, as discussed above, we discretize the continuous Euclidean space with the lattice spacing a:

$$x_\mu \to n_\mu a, \tag{3.15}$$

$$\phi(x) \to \phi(na), \tag{3.16}$$

$$\int d^4x \to a^4\sum_n, \tag{3.17}$$

$$\partial_\mu^2\phi(x) \to \frac{1}{a^2}\hat{\partial}^2\phi(na), \tag{3.18}$$

$$\mathcal{D}\phi \to \prod_n d\phi(na), \tag{3.19}$$

where $n_\mu = (n_1, n_2, n_3, n_4)$ and each component belongs to $\mathbb{Z}$. The derivative on the lattice is defined by the finite difference,

$$\hat{\partial}^2\phi(na) = \sum_\mu[\phi(na + \hat{\mu}a) + \phi(na - \hat{\mu}a) - 2\phi(na)], \tag{3.20}$$

with the unit vector $\hat{\mu} = \hat{e}_\mu$ along μ direction. With the use of a, whose dimension is inverse of mass, the field $\phi(na)$ and the mass are translated into the dimensionless one such that

$$\hat{\phi}_n = a\phi(na), \quad \hat{m}_0 = m_0 a. \tag{3.21}$$

Thus, we obtain the Green function on the lattice

$$\langle\hat{\phi}_{n^{(1)}}\cdots\hat{\phi}_{n^{(l)}}\rangle = \frac{\int\prod_m d\hat{\phi}_m\,\hat{\phi}_{n^{(1)}}\cdots\hat{\phi}_{n^{(l)}}e^{-S[\hat{\phi}]}}{\int\prod_m d\hat{\phi}_m\, e^{-S[\hat{\phi}]}}, \tag{3.22}$$

where the action expressed in terms of the dimensionless variables is

$$S[\hat{\phi}] = -\frac{1}{2}\sum_{n,\hat{\mu}} \hat{\phi}_n \hat{\phi}_{n+\hat{\mu}} + \frac{1}{2}(8+\hat{m}_0^2)\sum_n \hat{\phi}_n \hat{\phi}_n \equiv \frac{1}{2}\sum_{n,m} \hat{\phi}_n K_{nm} \hat{\phi}_m, \tag{3.23}$$

and the kernel apparently reads

$$K_{nm} = -\sum_{\mu>0}[\delta_{n+\hat{\mu},m} + \delta_{n-\hat{\mu},m} - 2\delta_{nm}] + \hat{m}_0^2 \delta_{nm}. \tag{3.24}$$

Since the action is originally dimensionless, there do not appear lattice spacings in the above expression.

Here, we comment on the choice of the action. The above expression of the action Eq. (3.23) is not unique but merely the simplest one, so that there is room for improvement. The only requirement which the lattice action should satisfy is that, in the continuum limit, it reduces to the correct action in the continuous theory. In addition, we choose the lattice action to keep the same symmetry that the action in continuous theories has as much as possible although there is explicit symmetry breaking on the lattice, such as the rotational symmetry and the translational symmetry. In the present case, the action for the scalar field theory on the lattice can be improved to embrace second-order and/or higher difference terms. In fact, while the scalar field is tractable in that the simple prescription leads to the correct action, the fermion field is hard to be handled as commented later.

Now, we have the action on the lattice and the path integral representations for it as well as the Green functions; we here consider the two-point function shortly as a specific example. The two-point function representing the propagator is obtained by a simple Gauss integral as

$$\langle \hat{\phi}_n \hat{\phi}_m \rangle = K^{-1}_{nm}, \tag{3.25}$$

where K^{-1} is the inverse matrix of K, defined as

$$\sum_l K_{nl} K^{-1}_{lm} = \delta_{nm}. \tag{3.26}$$

K^{-1} is explicitly written down as

$$K^{-1}_{nm} = \int_{-\pi}^{\pi} \frac{d^4\hat{k}}{(2\pi)^4} \frac{e^{i\hat{k}}(n-m)}{4\sum_\mu \sin^2(\hat{k}_\mu/2) + \hat{m}_0^2} \equiv G(n, m; \hat{m}_0). \tag{3.27}$$

In order to achieve the continuum limit of $G(n, m; \hat{m}_0)$, we take $a \to 0$ with physical quantities m_0, ϕ, and (x, y) fixed. In general, which quantity should be fixed when removing the lattice structure strongly depends on the theory one considers. For example, in an interacting theory, since the mass m_0 receives the non-trivial correction, the fixing of m_0 no longer makes sense. Nevertheless, in this free case, the naive prescription works, so that the correct two-point function in the continuous theory can be extracted by

$$\langle \phi(x)\phi(y) \rangle = \lim_{a \to 0} \frac{1}{a^2} G\left(\frac{x}{a}, \frac{y}{a}; m_0 a\right). \tag{3.28}$$

In fact, we confirm this from the relation,

$$G\left(\frac{x}{a}, \frac{y}{a}; m_0 a\right) = a^2 \int_{-\pi/a}^{\pi/a} \frac{d^4k}{(2\pi)^4} \frac{e^{ik(x-y)}}{\sum_\mu \tilde{k}_\mu^2 + m_0^2}, \tag{3.29}$$

where

$$\tilde{k}_\mu = \frac{2}{a} \sin \frac{k_\mu a}{2}. \tag{3.30}$$

The main contribution of the integral comes from the integration over the small momenta compared to $1/a$ because the integration is performed within $[-\pi/a, \pi/a]$. Thus setting $\tilde{k}_\mu \to k_\mu$ followed by $a \to 0$, we obtain the well-known result,

$$\langle \phi(x)\phi(y) \rangle = \int_{-\infty}^{\infty} \frac{d^4k}{(2\pi)^4} \frac{e^{ik\cdot(x-y)}}{k^2 + m_0^2}. \tag{3.31}$$

Since the above discussion is based on the fact that the lattice spacing is known before the study, we can take the continuum limit analytically. In actual numerical simulations, however, the lattice spacing does not explicitly appear and it is determined after all the calculations are accomplished. Moreover, in general, we do not have transcendental idea on the dependence of the observable on the lattice spacing. In principle, what we have to do is to make the lattice finer and finer. Then, when the correlation length $\hat{\xi} = \xi/a$ with the physical one ξ diverges or $\hat{m}_0 \to 0$, the continuum limit is achieved. The numerical implementation of such a setup is impossible.

Therefore, in order to study the continuum limit in the above case, we study the ratio $\langle \hat{\phi}_n \hat{\phi}_m \rangle / \hat{m}_0^2$ for small $\hat{m}_0$ on the fine lattice. For sufficiently small $\hat{m}_0$ that corresponds to the adequately fine lattice, the ratio becomes insensitive to $\hat{m}_0$. This means that the lattice we use is fine enough that we extract the continuum limit.

3.2.3 Free Fermion Field

In the free boson system, the simple action reduces to the correct continuum limit. In the case of the fermion field, however, the simple discretization of the well-known action,

$$S_{\mathrm{mat}}[\psi, \bar{\psi}] = \int d^4x \, \bar{\psi}(x)(\gamma_\mu \partial_\mu + m_0)\psi(x), \tag{3.32}$$

leads us to the notorious doubling problem, which states that there emerge unphysical degrees of freedom, called the doubler, in the continuum limit. We cannot distinguish those unphysical degrees from the genuine fermion at low energy. As a general statement, the Nielsen–Ninomiya theorem insists that there necessarily come up doublers if one uses the lattice formalism with all the following features [9]: hermiticity, locality, translational invariance, chiral symmetry, and bilinear form. Thus, in order to avoid the doubling problem, we have to break at least one of those conditions in the lattice formulation. There exist a lot of proposals, such as Wilson fermion [10], Staggered fermion (KS fermion) [11], Ginsparg–Wilson fermion (overlap fermion) [12], domain wall fermion [13], and so on. Since the object of this dissertation is not the theory including fermions but the pure YM theory, we do not go further into the discussion of those prescriptions. In the following, we tentatively employ the Wilson fermion for convenient discussion.

3.3 Lattice Gauge Theory

So far, we have overviewed the essential idea of the lattice field theory, taking examples of the boson and fermion field. While the actions discussed above have a global symmetry, they are not invariant under the local transformation depending on the space–time coordinate. The gauge principle requires the introduction of the gauge field. This fact is also true on the lattice. In this section, we discuss the gauge link and the gauge-invariant quantities constructed from link variables on the lattice.

3.3.1 Gauge Principle and Gauge Field

For simplicity, we take up the Abelian theory although the application to the non-Abelian theory is straightforward as shown later. The action for the free Dirac field is denoted by Eq. (3.32). We see that Eq. (3.32) has the global U(1) gauge symmetry under the transformation,

$$\psi(x) \to G\psi(x), \quad \bar{\psi}(x) \to \bar{\psi}(x)G^{-1}, \tag{3.33}$$

where $G = e^{i\alpha}$ with a constant parameter α. When we consider the local gauge transformation $\alpha = \alpha(x)$, then we have to introduce the covariant derivative,

$$D_\mu = \partial_\mu + A_\mu. \tag{3.34}$$

The modified action,

$$S_{\rm mat}[\psi, \bar{\psi}, A_\mu] = \int d^4x\, \bar{\psi}(x)(\gamma_\mu D_\mu + m_0)\psi(x), \tag{3.35}$$

remains invariant under the local gauge transformation,

$$\psi(x) \to G(x)\psi(x), \tag{3.36}$$

$$\bar{\psi}(x) \to \bar{\psi}(x)G^{-1}(x), \tag{3.37}$$

$$A_\mu(x) \to G(x)A_\mu(x)G^{-1}(x) - iG(x)\partial_\mu G^{-1}(x), \tag{3.38}$$

with $G = e^{i\alpha(x)}$. In the Abelian case, the transformation with respect to the gauge field reduces to $A_\mu(x) \to A_\mu(x) - \partial_\mu \alpha(x)$. Thus, the gauge principle is a strong statement that the local symmetry with the gauge field constrains the interaction between the matter field and the gauge field. This is true for all the interactions in the Standard Model as well.

Next, we need the kinetic term for the gauge field. The relativistic covariance and the gauge invariance lead to the kinetic term,

$$S_{\rm G} = \frac{1}{4e_0^2}\int d^4x\, F_{\mu\nu}(x)F_{\mu\nu}(x), \tag{3.39}$$

where the field strength $F_{\mu\nu}(x)$ is given by

$$F_{\mu\nu} = \partial_\mu A_\nu - \partial_\nu A_\mu, \tag{3.40}$$

and e_0 is the bare gauge coupling constant. In total, we obtain the well-known QED action,

$$S_{\rm QED} = \frac{1}{4e_0^2}\int d^4x\, F_{\mu\nu}F_{\mu\nu} + \int d^4x\, \bar{\psi}(x)(\gamma_\mu D_\mu + m_0)\psi(x). \tag{3.41}$$

3.3.2 *Link Variable*

The above discussion on the relation between the local gauge symmetry and the gauge field in the continuous space is also applicable to the gauge symmetry on the lattice. The essential point to study the gauge theory is to substitute the covariant derivative D_μ for the differential operator ∂_μ. On the lattice, we should handle the local transformation with the difference instead of the differential.

We start with the free Dirac action,

$$S_{\rm W} = (m_0 + 4r)\sum_n \bar{\psi}(n)\psi(n) - \frac{1}{2}\sum_{n,\mu}[\bar{\psi}(n)(r - \gamma_\mu)\psi(n+\hat{\mu}) + \bar{\psi}(n+\hat{\mu})(r + \gamma_\mu)\psi(n)], \tag{3.42}$$

where we employ the Wilson fermion with r being the Wilson parameter as the prescription to avoid the doubling problem. Although the fields are rendered dimensionless, we omit the hat notation $\hat{\cdot}$ except for the unit vector like $\hat{\mu}$ in the following. While the action, Eq. (3.42) is invariant under the global transformation,

$$\psi(n) \to G\psi(n), \quad \bar{\psi} \to \bar{\psi}(n)G^{-1}, \tag{3.43}$$

with $G \in$U(1), the local transformation $G(n)$ does not ensure the invariance of the second term in Eq. (3.42) due to the difference between coordinates n and $n + \hat{\mu}$.

In order to preserve the local gauge invariance, we introduce the link variable connecting between n and the neighboring site as

$$\bar{\psi}(n)(r - \gamma_\mu)\psi(n + \hat{\mu}) \to \bar{\psi}(n)(r - \gamma_\mu)U_{n,n+\hat{\mu}}\psi(n + \hat{\mu}), \tag{3.44}$$

$$\bar{\psi}(n + \hat{\mu})(r + \gamma_\mu)\psi(n) \to \bar{\psi}(n + \hat{\mu})(r + \gamma_\mu)U_{n+\hat{\mu},n}\psi(n), \tag{3.45}$$

where $U_{n+\hat{\mu},n} = U^\dagger_{n,n+\hat{\mu}} \in \mathrm{U}(1)$. The role of the link variable is to bridge the gauge transformations at different points just like the Wilson-line integral in the continuous theory. In particular, we write down $U_{n+\hat{\mu},n} = e^{i\phi_\mu(n)}$, where $\phi_\mu(n)$ is restricted within the compact region $[0, 2\pi]$. Since the link variable, or more correctly the phase $\phi(n)$, should reduce to the gauge field $A_\mu(x)$ in the continuum limit, we determine the following form,

$$U_\mu(n) \equiv U_{n,n+\hat{\mu}} = e^{iaA_\mu(n)}. \tag{3.46}$$

It is then understood that the corresponding gauge transformation is given by

$$\psi(n) \to G(n)\psi(n), \tag{3.47}$$

$$\bar{\psi}(n) \to \bar{\psi}(n)G^{-1}(n), \tag{3.48}$$

$$U_{n,n+\hat{\mu}} \to G(n)U_{n,n+\hat{\mu}}G^{-1}(n + \hat{\mu}), \tag{3.49}$$

$$U_{n+\hat{\mu},n} \to G(n + \hat{\mu})U_{n+\hat{\mu},n}G^{-1}(n), \tag{3.50}$$

which leads to the gauge covariant action,

$$S_{\mathrm{W}}[\psi, \bar{\psi}, U] = (\hat{M} + 4r)\sum_n \bar{\psi}(n)\psi(n) - \frac{1}{2}\sum_{n,\mu}[\bar{\psi}(n)(r - \gamma_\mu)U_{n,n+\hat{\mu}}\psi(n + \hat{\mu}) + \bar{\psi}(n + \hat{\mu})(r + \gamma_\mu)U^\dagger_{n+\hat{\mu}}\psi(n)]. \tag{3.51}$$

As shown in Fig. 3.1, the kinetic term of the gauge field can be represented with the plaquette,

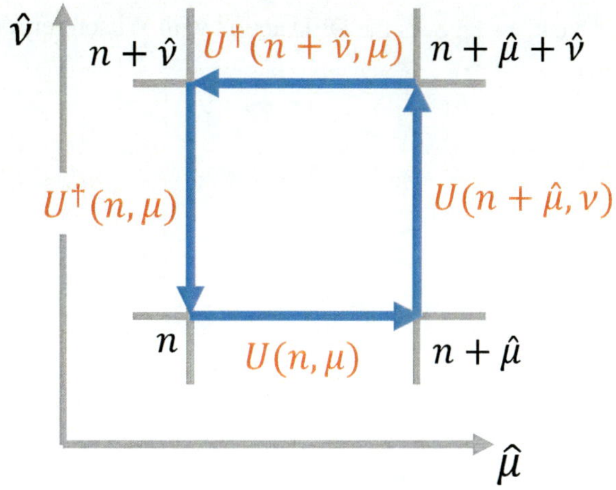

Fig. 3.1 Link variables and plaquette on the lattice. The closed loop ensures the gauge invariance

$$U_P \equiv U_{\mu\nu}(n) = U_\mu(n)U_\nu(n+\hat{\mu})U_\mu^\dagger(n+\nu)U_\nu^\dagger(n) = e^{ia^2 F_{\mu\nu}(n)}, \tag{3.52}$$

where the product follows the path order although this is not crucial in U(1) case. The field strength in the difference form is written down as

$$F_{\mu\nu}(n) = \frac{1}{a}[(A_\nu(n+\hat{\mu}) - A_\nu(n)) - (A_\mu(n+\hat{\nu}) - A_\mu(n))]. \tag{3.53}$$

The plaquette is clearly gauge invariant as well as scalar. Therefore, with the use of the plaquette, we construct the kinetic term as

$$S_{\mathrm{G}}[U] = \frac{1}{e_0^2}\sum_P \left[1 - \frac{1}{2}(U_P + U_P^\dagger)\right], \tag{3.54}$$

which correctly reproduces the kinetic term of the gauge field in the continuum limit.

As commented above in the case of scalar fields, the plaquette YM action is not a unique choice. In fact, whatever closed loops we construct are gauge invariant. As long as we appropriately choose a combination of closed loops to construct an action, respecting the correct space–time symmetry, the action is to reduce to the corresponding continuum action. Thus, we can design improved actions with sum of suitably averaged closed loops. The Symanzik action is one of the most famous improved actions, in which lattice artifacts are suppressed at finite lattice spacing. Other useful options are to employ a linear combination of 2×1 rectangles, chair-type, parallelogram loops, and/or bended Polyakov lines at next-to-leading order.

Now, we possess the QED action with Wilson fermion on the lattice,

$$
\begin{aligned}
S_{\mathrm{QED}}[\psi,\bar{\psi},U] =& \frac{1}{e_0^2}\sum_P\left[1-\frac{1}{2}(U_P+U_P^\dagger)\right] \\
&+(m_0+4r)\sum_n \bar{\psi}(n)\psi(n) \\
&-\frac{1}{2}\sum_{n,\mu}[\bar{\psi}(n)(r-\gamma_\mu)U_{n,n+\hat{\mu}}\psi(n+\hat{\mu})+\bar{\psi}(n+\hat{\mu})(r+\gamma_\mu)U_{n+\hat{\mu}}^\dagger\psi(n)].
\end{aligned}
\tag{3.55}
$$

3.3.3 Path Integral

We have introduced the classical action for the matter field and the gauge field, so that it be gauge invariant. This indicates that gauge-invariant observables are independent of the choice of the gauge. In the quantum theory on the basis of the path integral formalism, the gauge invariance is retained so that the integral measure be defined to be gauge invariant.

As is clear, the path integral is performed over all the variables, $\psi(n)$, $\bar{\psi}(n)$, and $U_\mu(n)$. For example, the multipoint correlation function is calculated by

$$
\begin{aligned}
&\langle \psi_\alpha(n)\cdots\bar{\psi}_\beta(m)\cdots U_\mu(l)\cdots\rangle \\
&=\frac{\int \mathcal{D}U\mathcal{D}\bar{\psi}\mathcal{D}\psi\;(\psi_\alpha(n)\cdots\bar{\psi}_\beta(m)\cdots U_\mu(l)\cdots)e^{-S_{\mathrm{QED}}[\psi,\bar{\psi},U]}}{\int \mathcal{D}U\mathcal{D}\bar{\psi}\mathcal{D}\psi\; e^{-S_{\mathrm{QED}}[\psi,\bar{\psi},U]}},
\end{aligned}
\tag{3.56}
$$

where the measure with respect to the link variable denotes

$$
\mathcal{D}U=\prod_{n,\mu} d\phi_\mu(n), \tag{3.57}
$$

which is referred to as the Haar measure, while the integration with respect to the matter fields is performed over the Grassmann variables. The Haar measure is invariant under any transformation in SU(N) group. One of the most remarkable features of the path integral on the lattice is that the integral is performed over the link variables, hence the integral over the compact group, which suggests that the integral is normalized. Thus, in the lattice gauge theory, we do not necessarily have to fix the gauge although the fixing of a suitable gauge is optional. The Elitzur theorem states that the expectation values of operators which are not gauge invariant are to be zero [14], which means that the local gauge symmetry is not spontaneously broken in the lattice gauge theory. In this sense, observables on the lattice are gauge-invariant quantities.

3.3.4 Non-Abelian Gauge Theory on the Lattice

The extension of the above formalism in U(1) theory into that in SU(N) gauge theory is rather straightforward. Now, the Dirac field possesses N components ψ^{α} labeled by $\alpha = 1, 2, \ldots, N$, and the gauge transformation G belongs to SU(N) group. The link variables are also represented as the adjoint representation,

$$U_{\mu}(n) = e^{i\phi_{\mu}(n)}, \tag{3.58}$$

with a Hermitian matrix $\phi_{\mu}(n) \in \mathfrak{su}(N)$.

While S_{W} in SU(N) does not receive a significant change except for the summation over the color indices, the kinetic term of the gauge field becomes the following form:

$$S_{\mathrm{G}}[U] = \beta \sum_{P} \left[1 - \frac{\mathrm{Tr}}{2N}(U_P + U_P^{\dagger}) \right]; \quad \beta = \frac{2N}{g_0^2}, \tag{3.59}$$

where the plaquette is now given by

$$U_P = e^{ia^2 G_{\mu\nu}(n)}, \tag{3.60}$$

with the field strength $G_{\mu\nu}(n)$. Here, the path ordering has to be kept due to the non-commutative nature.

In the almost same manner as the U(1) case, we can proceed to the quantum theory based on the path integral. Here, we restrict ourselves to the investigation of the SU(3) gauge theory, QCD, in particular. The correlation function, for instance, is given by

$$\langle \psi_{\alpha_1}^{a_1}(n_1) \cdots \bar{\psi}_{\beta_1}^{b_1}(m_1) \cdots U_{\mu_1}^{c_1 d_1}(k_1) \cdots \rangle \tag{3.61}$$

$$= \frac{\int \mathcal{D}U \mathcal{D}\bar{\psi} \mathcal{D}\psi \; \psi_{\alpha_1}^{a_1}(n_1) \cdots \bar{\psi}_{\beta_1}^{b_1}(m_1) \cdots U_{\mu_1}^{c_1 d_1}(k_1) \cdots e^{-S_{\mathrm{QCD}}}}{\int \mathcal{D}U \mathcal{D}\bar{\psi} \mathcal{D}\psi \; e^{-S_{\mathrm{QCD}}}}, \tag{3.62}$$

where the QCD action is

$$S_{\mathrm{QCD}} = S_{\mathrm{G}}[U] + S_{\mathrm{W}}[U, \psi, \bar{\psi}], \tag{3.63}$$

$$S_{\mathrm{G}} = \frac{6}{g_0^2} \sum_{P} \left[1 - \frac{\mathrm{Tr}}{6}(U_P + U_P^{\dagger}) \right], \tag{3.64}$$

$$S_{\mathrm{W}} = (\hat{M}_0 + 4r) \sum_{n} \bar{\psi}(n)\psi(n)$$
$$- \frac{1}{2} \sum_{n,\mu} [\bar{\psi}(n)(r - \gamma_{\mu}) U_{n,n+\hat{\mu}} \psi(n + \hat{\mu}) + \bar{\psi}(n + \hat{\mu})(r + \gamma_{\mu}) U_{n+\hat{\mu}}^{\dagger} \psi(n)]. \tag{3.65}$$

We notice that the integration measure is chosen to be the Haar measure.

3.4 Gauge-Invariant Quantity on the Lattice

So far, we concentrate on the construction of the action on the lattice and the method for the quantum theory. Now, we introduce specific observables on the lattice, which will be substantial ingredients for our studies in the later chapters.

3.4.1 *Wilson Loop*

As of today, there have been no observations that quarks and gluons behave isolatedly in nature. This is considered to be attributed to the self-interaction between gauge fields due to the non-Abelian aspect of QCD. This is referred to as the confinement, which is a truly non-perturbative phenomenon. There have been a lot of attempts to investigate the confinement by means of numerical simulations on the basis of the lattice field theory, where we prepare quark and anti-quark on the lattice. In vacuum, it is realized with Wilson loop.

The Wilson loop is one of the most fundamental observables on the lattice defined by

$$W_C[U] = \mathrm{Tr}\left[P\prod_{l\in C} U^{(l)}\right], \tag{3.66}$$

where $U^{(l)}$ represents the link variable belonging to the closed rectangular loop $C = R \times T$ shown in Fig. 3.2. R indicates the spatial distance, while T is the length along the imaginary time direction. The closed loop C ensures the gauge invariance of the Wilson loop just like a plaquette. On the T = const. plane, two fields whose representation in SU(N) is N and $\bar{N}$ reside at the distance R. The former representation corresponds to a static quark (Q) and the latter to a static anti-quark ($\bar{Q}$), whose positions are fixed with respect to the imaginary time. We notice that Q and $\bar{Q}$ are not dynamical quarks but test charges, so that they are independent of the dynamics

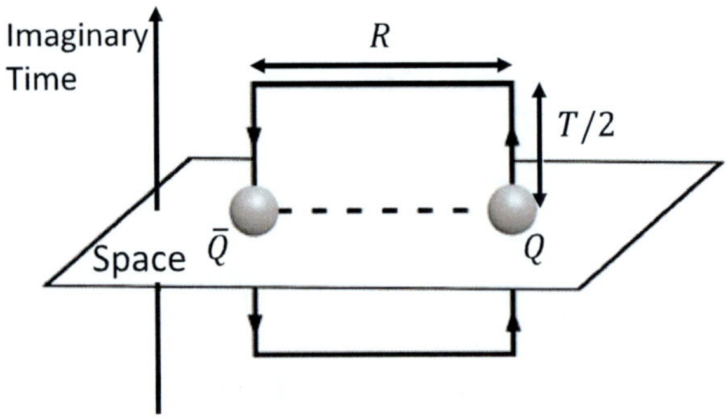

Fig. 3.2 Wilson loop whose spatial and temporal extension is R and T, respectively. This setup corresponds to the $Q\bar{Q}$ system at a distance R

of the theory. When we investigate the behavior of $W_C[U]$ in the $T \to \infty$ limit, we obtain

$$\langle W_C[U] \rangle \to e^{-TV_0(R)}, \tag{3.67}$$

where $V_0(R)$ is a static $Q\bar{Q}$ potential and the bracket represents the vacuum expectation value. In general, the Wilson loop contains the contribution from the ground state as well as the other higher excited states as

$$\langle W_C[U] \rangle = \sum_n C_n(R) e^{-TV_n(R)}, \tag{3.68}$$

with the overlaps $C_n(R)$ with $n = 0, 1, 2, \ldots$. Here, the limit $T \to \infty$ plays a role to reduce the contribution including C_n $(n \geq 1)$ and extract the ground state.

In actual numerical simulations, however, the implementation of $T \to \infty$ is not realistic and moreover it is difficult to extract the ground state because even the contribution of the ground state $C_0(R)e^{-TV_0(R)}$ damps exponentially. Therefore, we need to enhance the ground state that is the maximization of C_0 compared to C_n $(n \geq 1)$ by some smearing procedure such as APE.

3.4.2 Polyakov Loop

It has been confirmed that QCD experiences a non-trivial phase transition at a critical temperature T_c. To this effect, we can construct a gauge-invariant quantity with link variables on the lattice, which holds a non-trivial topology winding along the imaginary time direction as shown in Fig. 3.3. In the finite temperature lattice field theory, this is called the Polyakov loop, which can be constructed by virtue of the finite extension and the periodicity in the imaginary time direction.

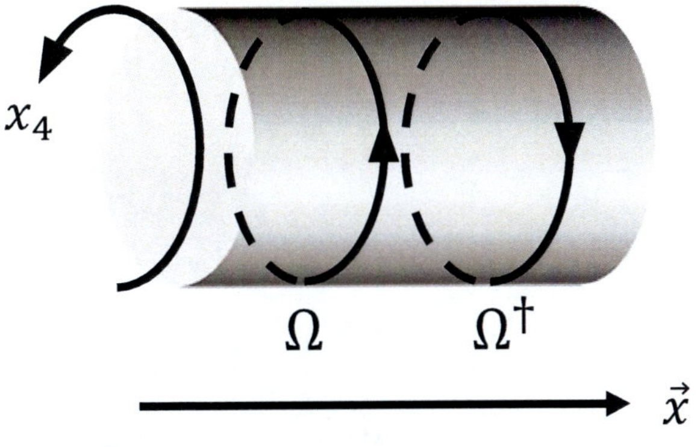

Fig. 3.3 Polyakov loop, which goes around the torus along the temporal direction

The Polyakov loop is defined as

$$\Omega(\mathbf{n}, T) = \mathrm{Tr}[U_4(\mathbf{n}, 1)U_4(\mathbf{n}, 2)\cdots U_4(\mathbf{n}, T)], \tag{3.69}$$

where the periodicity ensures the gauge invariance of $\Omega(\mathbf{n}, T)$. This quantity corresponds to the free energy necessary to create a single quark F_Q as follows:

$$\langle \Omega(\mathbf{n}, T) \rangle \propto e^{-TF_Q}. \tag{3.70}$$

In the confined phase, $\langle \Omega(\mathbf{n}, T) \rangle = 0$, while in the deconfined phase $\langle \Omega(\mathbf{n}, T) \rangle \neq 0$, hence the Polyakov loop plays a role of the order parameter to distinguish the confinement/deconfinement phase.

In fact, the pure lattice gauge theory has the following symmetry, which leads to $\langle \Omega(\mathbf{n}, T) \rangle = 0$. When we consider the global transformation independent of the spatial coordinate as

$$U_4(\mathbf{n}, T) \to G_t U_4(\mathbf{n}, T), \tag{3.71}$$

then the lattice action is invariant under this transformation. Here, $G_t \in Z_N$, where Z_N is a center of SU(N) group. Although, under this transformation, the product of link variables over the closed loop is invariant, the Polyakov loop is not invariant like

$$\Omega(\mathbf{n}, T) \to G_t \Omega(\mathbf{n}, T). \tag{3.72}$$

Thus, as far as Z_N symmetry is kept, the expectation value of the Polyakov loop vanishes, which means the Polyakov loop is eligible for an order parameter of Z_N symmetry. On the other hand, the spontaneous symmetry breaking of Z_N symmetry leads to the non-vanishing expectation value of the Polyakov loop. This phase transition is known to be first order, which is the same as that in the Z_N spin model, or the N-state Potts model. The existence and order of the phase transition are, indeed, established by the Monte Carlo simulations with the finite scaling analyses.

We notice that the above discussion is only applicable to the pure gauge theory. The existence of the dynamical fermion explicitly breaks the center symmetry, so that the expectation value of the Polyakov loop does not necessarily vanish even in the confined phase.

References

1. Wilson KG (1974) Confinement of quarks. Phys Rev D 10:2445
2. Creutz M (1979) Confinement and the critical dimensionality of space-time. Phys Rev Lett 43:553–556. [erratum: Phys. Rev. Lett. **43**, 890 (1979)]
3. Creutz M, Moriarty KJM (1982) Numerical studies of wilson loops in SU(3) gauge theory in four-dimensions. Phys Rev D 26:2166

4. Weingarten DH, Petcher DN (1981) Monte Carlo integration for lattice gauge theories with fermions. Phys Lett B 99:333–338
5. Rothe HJ (1992) Lattice gauge theories: an Introduction. World Sci Lect Notes Phys 43(1). [World Sci. Lect. Notes Phys. **59**, 1 (1997)] [World Sci. Lect. Notes Phys. **74**, 1 (2005)] [World Sci. Lect. Notes Phys. **82**, 1 (2012)]
6. Montvay I, Munster G Quantum fields on a lattice
7. DeGrand T, Detar CE, Lattice methods for quantum chromodynamics
8. Feynman RP (1948) Space-time approach to nonrelativistic quantum mechanics. Rev. Mod. Phys. 20:367
9. H. B. Nielsen and M. Ninomiya, "Absence of Neutrinos on a Lattice. 1. Proof by Homotopy Theory," Nucl. Phys. B **185**, 20 (1981) Erratum: [Nucl. Phys. B **195**, 541 (1982)]
10. Wilson KG, Quarks and strings on a lattice, CLNS-321
11. Kogut JB, Susskind L (1975) Hamiltonian formulation of wilson's lattice gauge theories. Phys Rev D 11:395
12. Ginsparg PH, Wilson KG (1982) A Remnant of Chiral Symmetry on the Lattice. Phys. Rev. D 25:2649
13. Kaplan DB (1992) A Method for simulating chiral fermions on the lattice. Phys Lett B 288:342–347. arXiv:hep-lat/9206013 [hep-lat]
14. Elitzur S (1975) Impossibility of Spontaneously Breaking Local Symmetries. Phys. Rev. D 12:3978

Chapter 4
Yang–Mills Gradient Flow and Energy Momentum Tensor

In this chapter, we recapitulate the features of the Yang–Mills (YM) gradient flow [1–3] and the method to define the renormalized energy momentum tensor (EMT) by means of it [4, 5]. The gradient flow is one of the smearing methods to suppress short-distance or ultraviolet fluctuations of quantum field configurations. A remarkable feature is that it is a continuous transformation in contrast to discrete ones, such as APE [6], STOUT [7] and HYP smearings [8] with weighted link staples.

The gradient flow was proposed to be an effective solution to solve the critical slowing down, which is the fact that the computational cost around the critical points of a theory conspicuously increases due to the long autocorrelation time [9]. The rapid growth of the autocorrelation time results from the fixing of the topological charge when the continuum limit is approached. This fact prevents us from correctly evaluating the statistical and systematic errors in the lattice simulations based on the Monte Carlo methods. Reference [10] originally proposed the use of the trivializing map with the hybrid Monte Carlo algorism with the use of the gradient flow, which contributes to the reduction of the autocorrelation time since this map makes the theory closer to the strong coupling region. The numerical test was performed in, for instance, Ref. [11].

The gradient flow itself and its applications today gather attentions, and there have been a lot of studies with the gradient flow, such as the scale setting, the definition of the topological charge and so forth [2, 12, 13]. The gradient flow also has a remarkable characteristic in defining the UV finite quantities even for the local operator products, one of whose properties leads us to the construction of EMT. Recently, in order to confirm the validity of the use of the EMT operator via the gradient flow, the equation of state in the SU(3) YM gauge theory at finite temperature has been studied with this method [14–16]. It has been confirmed that the results are consistent with those in the other works with the integral method and so on [17–20]. The usefulness of the method has been also found in that it gives small noise with small statistics compared with the previous studies with high statistics.

R. Yanagihara, *Distribution of Energy Momentum Tensor around Static Charges in Lattice Simulations and an Effective Model*, Springer Theses,
https://doi.org/10.1007/978-981-16-6234-8_4

This chapter is organized in the following manner. We first give a brief review on the YM theory and its renormalizability in Sect. 4.1 based on Refs. [2, 3, 21]. The explanation of the YM gradient flow in Sect. 4.2 follows by the proof of the renormalizability of the gradient flow in Sect. 4.3. The definition of the EMT operator on the basis of the method is given in Sect. 4.4 [4], where we comment on the double extrapolation to obtain physical quantities especially in actual simulations. The final Sect. 4.5 is dedicated to remarks on fermions [5].

4.1 Quantum Theory of Yang–Mills Theory

4.1.1 Action

The theory to describe the dynamics of the elementary particles is known as 'the Standard Model.' In our studies, we focus on the theory of the strong interaction among quarks and gluons, that is Quantum Chromodynamics (QCD). It has a local color SU(3) symmetry, whose action is given by

$$\begin{aligned} S &= S_{\mathrm{mat}} + S_{\mathrm{G}} \\ &= \int d^D x \sum_{f=\mathrm{flavors}} \bar{\psi}_f^{\alpha}(x)(\not{D}_{\alpha\beta} + m_0^f \delta_{\alpha\beta})\psi_f^{\beta}(x) + \frac{1}{4g_0^2}\int d^D x\, G_{\mu\nu}^a(x) G_{\mu\nu}^a(x), \end{aligned} \tag{4.1}$$

where S_{mat} represents the matter part including the quark field $\psi_f^{\alpha}(x)$ and its interaction with gluons while S_{G} the kinetic term of the gauge fields. In S_{mat}, $\psi_f^{\alpha}(x)$ and $\bar{\psi}_f^{\alpha}(x)$, belonging to the fundamental representation in the SU(3) Lie group, denote the quark field, whose bare mass is m_0^f with respect to six flavors labeled by f. The indices α, β represent the color degrees of freedom, which run from 1 to 3. The covariant derivative D_μ is defined as

$$D_\mu = \partial_\mu + A_\mu^a(x) T^a, \tag{4.2}$$

with the bare gauge field $A_\mu^a(x)$ and Gell–Mann matrix T^a, whose index a runs 1 through 8. The existence of the covariant derivative guarantees the local gauge invariance. In S_{G}, the field strength $G_{\mu\nu}^a(x)$ $(\mu, \nu = 1, 2, 3, 4$ and $a = 1, 2, \ldots, 8)$ composed of $A_\mu^a(x)$ is

$$G_{\mu\nu}^a = \partial_\mu A_\nu^a(x) - \partial_\nu A_\mu^a(x) + f^{abc} A_\mu^b(x) A_\nu^c(x), \tag{4.3}$$

with the SU(3) structure constant f^{abc} and g_0 represents the bare gauge coupling. The last term in Eq. (4.3) gives rise to the nonlinear self-interaction between the gauge particles, so that it is exceedingly difficult to analyze QCD. Throughout this dissertation, we mainly consider the pure SU(3) YM gauge theory in D-dimensional Euclidean space given by $S_{\rm G}$, where $D = 4 - 2\epsilon$. In other words, the fermions have infinitely heavy mass so that the fermion determinant be constant. This is called the quenched QCD.

4.1.2 Quantization

The classical action $S_{\rm G}$ and the equation of motion derived from it are not enough to understand the real dynamics of Yang–Mills theory since the quantum effect plays a crucial role. In this section, we consider the quantization of the YM theory.

While there are several manners to quantize the gauge theories, we employ the quantization on the basis of the Feynman path integral method. In the path integral, we define the partition function $\mathcal{Z}$ as

$$\mathcal{Z} = \int \mathcal{D}A \, e^{-S_{\rm G}}. \tag{4.4}$$

As is commented above, this is a gauge theory so that the action $S_{\rm G}$ is invariant under the following infinitesimal gauge transformation,

$$\delta A^a_\mu(x) = \partial_\mu w^a(x) + f^{abc} A^b_\mu(x) w^c(x) \equiv D_\mu w^a(x), \tag{4.5}$$

with some scalar function $w^a(x)$. The gauge invariance prevents a simple perturbative expansion of Eq. (4.4) because of the multiple counting with respect to the infinite number of physically equivalent fields.[1] In order to develop the well-defined perturbation theory, we need a gauge fixing by adding the gauge fixing term,

$$S_{\rm gf} = \frac{\lambda_0}{2g_0^2} \int d^D x \, (\partial_\mu A^a_\mu(x))^2, \tag{4.6}$$

to the action $S_{\rm G}$. $S_{\rm gf}$ suggests the emergence of the Faddeev–Popov ghost–anti-ghost term $S_{c\bar{c}}$. Thus, the required terms are summarized in the following form:

$$S_{\rm gf} + S_{c\bar{c}} = \frac{1}{g_0^2} \int d^D x \left(B^a(x) \partial_\mu A^a_\mu(x) - \frac{1}{2\lambda_0} B^a(x) B^a(x) + \frac{1}{2\lambda_0} \bar{c}^a(x) \partial_\mu D_\mu c^a(x) \right). \tag{4.7}$$

Here we employ the auxiliary field, so-called the Nakanishi–Lautrup (NL) field $B^a(x)$ for the purpose of later convenience. In fact, when we eliminate the NL field by the equation of motion,

[1] Note that the gauge fixing is not necessary when we numerically perform the functional integrations thanks to the integration over the compact group manifold.

$$B^a(x) = \lambda_0 \partial_\mu A^a_\mu(x), \tag{4.8}$$

the gauge fixing term is reproduced. The total action can be written as

$$S_{\text{gf}} + S_{c\bar{c}} = \delta \frac{1}{g_0^2} \int d^D x\, \bar{c}^a(x) \left(\partial_\mu A^a_\mu(x) - \frac{1}{2\lambda_0} B^a(x) \right), \tag{4.9}$$

where δ denotes the nilpotent Becchi–Rouet–Stora (BRS) transformation defined by

$$\delta A^a_\mu(x) = D_\mu c^a(x), \tag{4.10}$$

$$\delta c^a(x) = -\frac{1}{2} f^{abc} c^b(x) c^c(x), \tag{4.11}$$

$$\delta \bar{c}^a(x) = B^a(x), \tag{4.12}$$

$$\delta B^a(x) = 0. \tag{4.13}$$

$S_{\text{gf}} + S_{c\bar{c}}$ is BRS invariant, so as for S_{G}. This fact indicates that any gauge-invariant operators are independent of the gauge fixing parameter λ_0, so that λ_0 has no effect on the physical quantities.

4.1.3 One-Loop Divergences

Here we compute the Feynman diagrams containing the YM gauge vector particles up to one-loop order and find the UV divergences of the YM theory. We now have the action $S_{\text{G}} + S_{\text{gf}} + S_{c\bar{c}}$. Thus, the free propagators for the gauge field and the ghost field are calculated from functional integrals of two-point functions over $A^a_\mu(x)$ and $c^a(x)$, respectively. The results are expressed by

$$\langle A^a_\mu(x) A^b_\nu(y) \rangle_0 = g_0^2 \delta^{ab} \int_p \frac{e^{ip(x-y)}}{(p^2)^2} \left[(\delta_{\mu\nu} p^2 - p_\mu p_\nu) + \frac{1}{\lambda_0} p_\mu p_\nu \right], \tag{4.14}$$

$$\langle c^a(x) c^b(y) \rangle_0 = g_0^2 \delta^{ab} \int_p \frac{e^{ip(x-y)}}{p^2}, \tag{4.15}$$

where we use a shorthand notation for the momentum integration in D dimension,

$$\int_p \equiv \int \frac{d^D p}{(2\pi)^D}. \tag{4.16}$$

Now we proceed to the computation of the divergent diagrams in the one-loop level for the renormalization discussed later. In what follows, we employ the Feynman gauge $\lambda_0 = 1$ for the computational simplicity. First, we consider the ghost two-point function. The only contribution to $\langle c^a(x) \bar{c}^b(y) \rangle$ is Fig. 4.1. Note that the tadpole

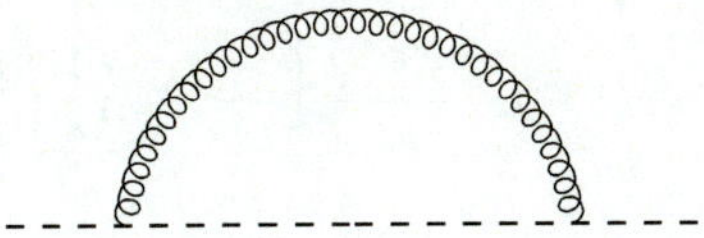

Fig. 4.1 Two-point function of the ghost propagator at one-loop order. The spiral (dashed) line represents the gluon (ghost) propagator

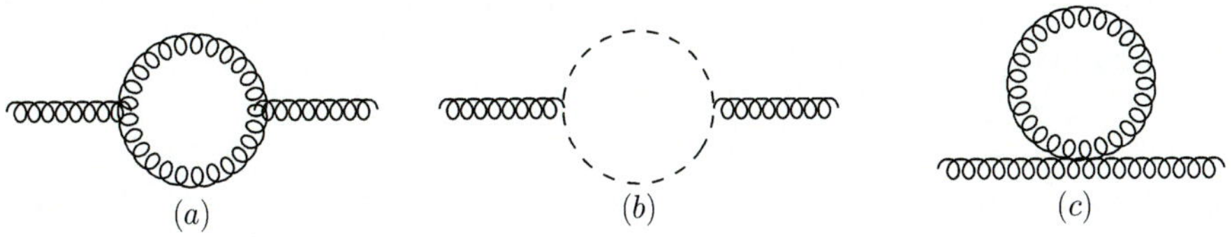

Fig. 4.2 Two-point function of gauge fields

diagram vanishes because δ^{ab} contracts the indices of the structure function. After straightforward calculation, we obtain[2]

$$\langle c^a(x)\bar{c}^b(y)\rangle = \frac{g_0^4}{(4\pi)^2}C_A\delta^{ab}\int_p \frac{e^{ip(x-y)}}{p^2}\int_0^1 dx\, x\left[\frac{1}{\epsilon} - \gamma_E - \ln\left(\frac{x(1-x)p^2}{4\pi}\right)\right], \tag{4.17}$$

where C_A represents the Casimir operator and in SU(N) case $C_A = N$. Therefore, we find that the divergence is given by

$$\langle c^a(x)\bar{c}^b(y)\rangle_{\text{1-loop, div.}} = g_0^2\delta^{ab}\int_p \frac{e^{ip(x-y)}}{p^2}\left[\frac{g_0^2}{(4\pi)^2}\frac{1}{2}C_A\frac{1}{\epsilon}p^2\right]. \tag{4.18}$$

Second, we calculate $\langle A_\mu^a(x)A_\nu^b(y)\rangle$ whose contributions are shown in Fig. 4.2. The diagram (b) in Fig. 4.2 includes the ghost loop. Simple calculations including the integration over the Feynman parameter provide us the divergent part in $\langle A_\mu^a(x)A_\nu^b(y)\rangle$,

[2] Throughout the perturbative calculation, we employ the dimensional regularization while soon after we consider the lattice regularization. In the dimensional regularization, we have the following formula,

$$\int_p \frac{1}{(p^2+M^2)^\alpha} = \frac{\Gamma(\alpha - D/2)}{(4\pi)^{D/2}\Gamma(\alpha)}M^{D-2\alpha},$$

with parameters M and α. $\Gamma(x)$ denotes the Gamma function.

$$\langle A_\mu^a(x)A_\nu^b(y)\rangle_{\text{(a), 1-loop, div.}} = g_0^2\delta^{ab}\int_p \frac{e^{ip(x-y)}}{p^2}\left[\frac{g_0^2}{(4\pi)^2}C_A\frac{1}{\epsilon}\left(\frac{19}{12}\delta_{\mu\nu}p^2 - \frac{11}{6}p_\mu p_\nu\right)\right], \tag{4.19}$$

$$\langle A_\mu^a(x)A_\nu^b(y)\rangle_{\text{(b), 1-loop, div.}} = g_0^2\delta^{ab}\int_p \frac{e^{ip(x-y)}}{p^2}\left[\frac{g_0^2}{(4\pi)^2}C_A\frac{1}{\epsilon}\left(\frac{1}{12}\delta_{\mu\nu}p^2 + \frac{1}{6}p_\mu p_\nu\right)\right], \tag{4.20}$$

$$\langle A_\mu^a(x)A_\nu^b(y)\rangle_{\text{(c), 1-loop, div.}} = 0, \tag{4.21}$$

where the subscripts (a), (b) and (c) mean each diagram in Fig. 4.2. The third diagram (c) vanishes under the dimensional regularization. Thus, the sum of divergent parts of the above diagrams results in the following form:

$$\langle A_\mu^a(x)A_\nu^b(y)\rangle_{\text{1-loop, div.}} = g_0^2\delta^{ab}\int_p \frac{e^{ip(x-y)}}{p^2}\left[\frac{g_0^2}{(4\pi)^2}C_A\frac{1}{\epsilon}\frac{5}{3}(\delta_{\mu\nu}p^2 - p_\mu p_\nu)\right]. \tag{4.22}$$

Finally, we consider ghost–anti-ghost-gauge three-point function shown in Fig. 4.3. Up to the one-loop order, there are two one-particle irreducible (1PI) diagrams, Fig. 4.3(a) and (b) as well as four one-particle reducible (1PR) diagrams, Fig. 4.3(c)–(f). The first 1PI diagram containing 'AAA' vertex yields

$$\begin{aligned}&\langle c^a(x)A_\mu^b(y)\bar{c}^c(z)\rangle_{\text{(a), 1-loop, div.}}\\ &= g_0^4 f^{abc}\int_{p,q} e^{ip(x-y)}e^{ip(y-z)}\frac{1}{p^2}\frac{1}{(p+q)^2}\frac{1}{q^2}\left[\frac{g_0^2}{(4\pi)^2}\frac{1}{\epsilon}\frac{3}{8}C_A\right]ip_\mu,\end{aligned} \tag{4.23}$$

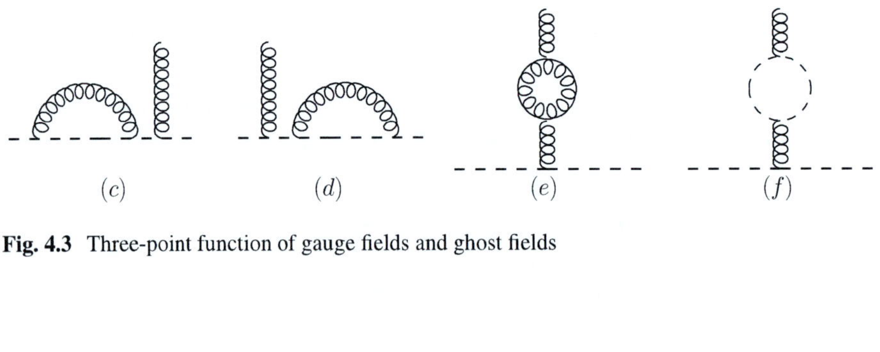

Fig. 4.3 Three-point function of gauge fields and ghost fields

where we use the identity

$$f^{aXb}f^{bYc}f^{cZa} = -\frac{1}{2}f^{XYZ}C_A. \tag{4.24}$$

Then the other 1PI diagram with three '$\bar{c}Ac$' vertices gives

$$\begin{aligned}&\langle c^a(x)A^b_\mu(y)\bar{c}^c(z)\rangle_{\text{(b), 1-loop, div.}}\\ =&g_0^4 f^{abc}\int_{p,q} e^{ip(x-y)}e^{ip(y-z)}\frac{1}{p^2}\frac{1}{(p+q)^2}\frac{1}{q^2}\left[\frac{g_0^2}{(4\pi)^2}\frac{1}{\epsilon}\frac{1}{8}C_A\right]ip_\mu.\end{aligned} \tag{4.25}$$

We also have four 1PR diagrams, resulting in

$$\begin{aligned}&\langle c^a(x)A^b_\mu(y)\bar{c}^c(z)\rangle_{\text{(c-f), 1-loop, div.}}\\ =&g_0^4 f^{abc}\int_{p,q} e^{ip(x-y)}e^{ip(y-z)}\frac{1}{p^2}\frac{1}{(p+q)^2}\frac{1}{q^2}\\ &\times\left[\frac{g_0^2}{(4\pi)^2}\frac{1}{\epsilon}C_A\left\{\left(\frac{5}{3}+2\cdot\frac{1}{2}\right)\delta_{\mu\nu}q^2-\frac{5}{3}q_\mu q_\nu\right\}\right]ip_\nu.\end{aligned} \tag{4.26}$$

In total, we find that the structure of the divergence in the ghost–anti-ghost-gauge three-point function is represented by

$$\begin{aligned}&\langle c^a(x)A^b_\mu(y)\bar{c}^c(z)\rangle_{\text{1-loop, div.}}\\ =&g_0^4 f^{abc}\int_{p,q} e^{ip(x-y)}e^{ip(y-z)}\frac{1}{p^2}\frac{1}{(p+q)^2}\frac{1}{q^2}\\ &\times\left[\frac{g_0^2}{(4\pi)^2}\frac{1}{\epsilon}C_A\left(\frac{19}{6}\delta_{\mu\nu}q^2-\frac{5}{3}q_\mu q_\nu\right)\right]ip_\nu.\end{aligned} \tag{4.27}$$

4.1.4 Renormalization

As in quantum electrodynamics, non-Abelian gauge symmetry restricts the structure of divergences. In this section, using the result obtained in the previous section, we prove that all the UV divergences in correlation functions are absorbed into only three divergent constants $Z,\ Z_3,\ \tilde{Z}_3$:

$$g_0^2 = \mu^{2\epsilon} g^2 Z, \tag{4.28}$$

$$\lambda_0 = \lambda Z_3^{-1}, \tag{4.29}$$

$$A_\mu^a(x) = Z^{1/2} Z_3^{1/2} (A_{\rm R})_\mu^a(x), \tag{4.30}$$

$$c^a(x) = \tilde{Z}_3 Z^{1/2} Z_3^{1/2} c_{\rm R}^a(x), \tag{4.31}$$

$$\bar{c}^a(x) = Z^{1/2} Z_3^{-1/2} \bar{c}_{\rm R}^a(x), \tag{4.32}$$

order by order in the perturbation theory, where

$$Z_i = 1 + Z_i^{(1)} + Z_i^{(2)} + \cdots \quad (Z_i = Z,\ Z_3,\ \tilde{Z}_3). \tag{4.33}$$

In particular, $Z_i^{(1)}$ acquire the divergence proportional to $1/\epsilon$. In the following, we discuss the one-loop-order renormalization.

To the one-loop order,

$$\begin{aligned}
&\langle (A_{\rm R})_\mu^a(x)(A_{\rm R})_\nu^b(y)\rangle \\
=&Z^{-1}Z_3^{-1}[\langle A_\mu^a(x)A_\nu^b(y)\rangle_{\text{tree}} + \langle A_\mu^a(x)A_\nu^b(y)\rangle_{\text{1-loop}}] \\
=&\mu^{2\epsilon} g^2 \delta^{ab} \int_p \frac{e^{ip(x-y)}}{(p^2)^2} \left[Z_3^{-1}(\delta_{\mu\nu}p^2 - p_\mu p_\nu) + \frac{1}{\lambda} p_\mu p_\nu \right] + \langle A_\mu^a(x)A_\nu^b(y)\rangle_{\text{1-loop}}.
\end{aligned} \tag{4.34}$$

Therefore, by choosing

$$Z_3^{(1)} = \frac{g^2}{(4\pi)^2} \frac{1}{\epsilon} \frac{5}{3} C_A, \tag{4.35}$$

the one-loop divergence in the correlation function $\langle A_\mu^a(x)A_\nu^b(y)\rangle$ can be removed with the minimal subtraction scheme. We notice that, in the one-loop-order calculation, we regard $g_0 = g$.

Similarly, for the determination of $\tilde{Z}_3^{(1)}$,

$$\begin{aligned}
&\langle c_{\rm R}^a(x)\bar{c}_{\rm R}^b(y)\rangle \\
=&\tilde{Z}_3^{-1}Z^{-1}[\langle c^a(x)\bar{c}^b(y)\rangle_{\text{tree}} + \langle c^a(x)\bar{c}^b(y)\rangle_{\text{1-loop}}] \\
=&\mu^{2\epsilon} g^2 \delta^{ab} \int_p \frac{e^{ip(x-y)}}{p^2} \tilde{Z}_3^{-1} + \langle c^a(x)\bar{c}^b(y)\rangle_{\text{1-loop}},
\end{aligned} \tag{4.36}$$

hence

$$\tilde{Z}_3^{(1)} = \frac{g^2}{(4\pi)^2} \frac{1}{\epsilon} \frac{1}{2} C_A. \tag{4.37}$$

The last constant Z can be deduced from the three-point function as follows:

$$\begin{aligned}
&\langle c_{\mathrm{R}}^a(x)(A_{\mathrm{R}})_\mu^b(y)\bar{c}_{\mathrm{R}}^c(z)\rangle \\
=&\tilde{Z}_3^{-1}Z^{-3/2}Z_3^{-1/2}[\langle c^a(x)A_\mu^b(y)\bar{c}^c(z)\rangle_{\text{tree}} + \langle c^a(x)A_\mu^b(y)\bar{c}^c(z)\rangle_{\text{1-loop}}] \\
=&\mu^{4\epsilon}g^4 f^{abc}\int_{p,q} e^{ip(x-y)}e^{iq(y-z)}\frac{1}{p^2}\frac{1}{(p+q)^2}\frac{1}{q^2} \\
&\times\left[\tilde{Z}_3^{-1}Z^{1/2}Z_3^{-1/2}(\delta_{\mu\nu}q^2 - q_\mu q_\nu) + \tilde{Z}_3^{-1}Z^{1/2}Z_3^{1/2}\frac{1}{\lambda}q_\mu q_\nu\right]ip_\nu \\
&+ \langle c^a(x)A_\mu^b(y)\bar{c}^c(z)\rangle_{\text{1-loop}}.
\end{aligned} \tag{4.38}$$

The cancellation of the divergence requires

$$Z^{(1)} = \frac{g^2}{(4\pi)^2}\frac{1}{\epsilon}\left(-\frac{11}{3}\right)C_A. \tag{4.39}$$

Now let us summarize the one-loop-order renormalization in the Feynman gauge:

$$g_0^2 = \mu^{2\epsilon}g^2\left[1 + \frac{g^2}{(4\pi)^2}\frac{1}{\epsilon}\left(-\frac{11}{3}\right)C_A\right], \tag{4.40}$$

$$\lambda_0 = \lambda\left[1 + \frac{g^2}{(4\pi)^2}\frac{1}{\epsilon}\left(-\frac{5}{3}\right)C_A\right], \tag{4.41}$$

$$A_\mu^a(x) = \left[1 + \frac{g^2}{(4\pi)^2}\frac{1}{\epsilon}(-1)C_A\right](A_{\mathrm{R}})_\mu^a(x), \tag{4.42}$$

$$c^a(x) = \left[1 + \frac{g^2}{(4\pi)^2}\frac{1}{\epsilon}\left(-\frac{1}{2}\right)C_A\right]c_{\mathrm{R}}^a(x), \tag{4.43}$$

$$\bar{c}^a(x) = \left[1 + \frac{g^2}{(4\pi)^2}\frac{1}{\epsilon}\left(-\frac{8}{3}\right)C_A\right]\bar{c}_{\mathrm{R}}^a(x). \tag{4.44}$$

We finally note that, in the modified minimal subtraction ($\overline{\mathrm{MS}}$) scheme, we simply replace the renormalization scale μ^2 by $4\pi e^{-\gamma_E}\mu^2$.

4.2 Introduction to Gradient Flow

4.2.1 *Idea of Yang–Mills Gradient Flow*

Now we introduce the gradient flow. The YM gradient flow is the transformation of the gauge fields $A_\mu^a(x)$ into the flowed gauge field $B_\mu^a(t,x)$ with the fictitious fifth dimension, called the flow time $t \in [0,\infty)$ newly introduced in addition to the ordinary four-dimensional Euclidean space $x \in \mathbb{R}^4$. The evolution $A_\mu^a(x) \to$

$B^a_\mu(t, x)$ along the flow time direction is performed according to the deterministic YM equation of motion, named the flow equation:

$$\partial_t B^a_\mu(t, x) = -g_0^2 \frac{\delta S_G(t)}{\delta A^a_\mu(t, x)} = D_\nu G^a_{\nu\mu}(t), \tag{4.45}$$

where $S_G(t)$ and $G^a_{\mu\nu}(t, x)$ represent the YM action and the flowed field strength with the replacement of $A^a_\mu(x)$ by $B^a_\mu(t, x)$, respectively. We emphasize that x belongs to $\mathbb{R}^4$ and the imaginary time x_4 have nothing to do with the flow time t. Note that the physical dimension of t is $[\mathrm{M}^{-2}]$. The initial value is set to be the conventional gauge field,

$$B^a_\mu(t = 0, x) = A^a_\mu(x). \tag{4.46}$$

The right-hand side of Eq. (4.45) is the gradient in functional space if $S_G(t)$ is regarded as a potential function, hence we call this equation as 'gradient flow.' We can regard the flow equation as a sort of diffusion equation. In fact, when we expand the right-hand side of Eq. (4.45) up to the first order of the flowed gauge field, we obtain

$$\partial_t B^a_\mu(t, x) = \partial^2 B^a_\mu(t, x) + \cdots, \tag{4.47}$$

where the diffusion length is $\sim\sqrt{2t}$. Thus, the flow $t > 0$ makes the field configurations smooth.

Similarly, in the lattice gauge theory, we can formulate the gradient flow. The flow equation for the link variable $V_\mu(t, x)$ is written by

$$\partial_t V_\mu(t, x) = -g_0^2 \frac{\delta S(t)}{\delta V_\mu(t, x)}, \tag{4.48}$$

whose initial condition is

$$V_\mu(t = 0, x) = U_\mu(x), \tag{4.49}$$

where $U_\mu(x)$ denotes the ordinary link variable in the D-dimensional space. $S(t)$ represents the action constructed with the flowed gauge field. We employ the Wilson gauge action

$$S_W[U] = \frac{1}{g_0^2} \sum_P \mathrm{Re}\,\mathrm{Tr}[1 - U_P], \tag{4.50}$$

where U_P indicates the plaquette.

4.2.2 Perturbative Expansion of Gradient Flow

In order to suppress the gauge modes, in the following, we employ the modified flow equation to which modification terms are added as follows:

$$\partial_t B_\mu(t,x) = D_\nu G_{\nu\mu}(t,x) + \alpha_0 D_\mu \partial_\nu B_\nu(t,x). \tag{4.51}$$

Gauge-invariant quantities do not depend on the gauge fixing parameter α_0. Under the infinitesimal gauge transformation,

$$B^a_\mu(t,x) \to B^a_\mu(t,x) + D_\mu w^a(t,x), \tag{4.52}$$

the flow equation reduces to

$$\partial_t B^a_\mu(t,x) = D_\nu(t,x) G^a_{\nu\mu}(t,x) + \alpha_0 D_\mu \partial_\nu B^a_\nu(t,x) - D_\mu(\partial_t - \alpha_0 D_\nu \partial_\nu) w^a(t,x). \tag{4.53}$$

Therefore, by choosing $w^a(t,x)$ such that

$$(\partial_t - \alpha_0 D_\nu \partial_\nu) w^a(t,x) = -\delta\alpha_0 \partial_\nu B^a_\nu(t,x) \;\; ; \;\; w^a(t=0,x) = 0, \tag{4.54}$$

α_0 can be changed according to

$$\alpha_0 \to \alpha_0 + \delta\alpha_0. \tag{4.55}$$

This means that given parameter α_0 corresponds to one gauge field $B^a_\mu(t,x)$, so that the gauge fields themselves are connected with one another via the gauge transformation. Thus, any gauge-invariant quantities are independent of α_0 as far as it does not contain the derivative with respect to t.

The flow equation (4.51) can be formally solved as

$$B^a_\mu(t,x) = \int d^D y \left[K_t(x-y)_{\mu\nu} A^a_\nu(y) + \int_0^t ds\, K_{t-s}(x-y)_{\mu\nu} R^a_\nu(s,y) \right], \tag{4.56}$$

with the heat kernel,

$$K_t(x)_{\mu\nu} = \int_p \frac{e^{ipx}}{p^2} [(\delta_{\mu\nu} p^2 - p_\mu p_\nu) e^{-tp^2} + p_\mu p_\nu e^{-\alpha_0 t p^2}]. \tag{4.57}$$

Here $R^a_\mu(t,x)$ represents the nonlinear terms in the flow equation,

$$R^a_\mu = f^{abc}[2B^b_\nu \partial_\nu B^c_\mu - B^b_\nu \partial_\mu B^c_\nu + (\alpha_0 - 1) B^b_\mu \partial_\nu B^c_\nu] + f^{abc} f^{cde} B^b_\nu B^d_\nu B^e_\mu. \tag{4.58}$$

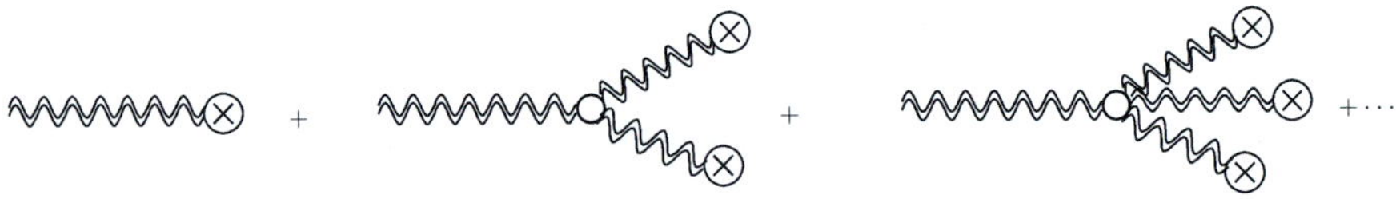

Fig. 4.4 Diagrammatic expression of the solution Eq. (4.56). The double wavy line represents the heat kernel $K_t(x)_{\mu\nu}$. The crosses and the white circles are $A^a_\mu(x)$ and $R^a_\mu(t,x)$, respectively

In particular, up to the first order with respect to the gauge field, the formal solution, Eq. (4.56) reads

$$B^a_\mu(t,x) = \int d^D y\, K_t(x-y)_{\mu\nu} A^a_\nu(y), \tag{4.59}$$

where the kernel is

$$K_t(x)_{\mu\nu} = \delta_{\mu\nu} \int_p e^{ipx} e^{-tp^2} = \frac{e^{-x^2/4t}}{(4\pi t)^{D/2}}. \tag{4.60}$$

Here we set $\alpha_0 = 1$ for a plain explanation. We can confirm that the gauge field is smeared within the radius $\sqrt{2t}$ in the D-dimensional Euclidean space. Equation (4.56) can be iteratively solved and the solution is then represented diagrammatically as shown in Fig. 4.4, where the double lines, crosses and white circles indicate $K_t(x)_{\mu\nu}$, $A^a_\mu(x)$ and $R^a_\mu(t,x)$, respectively.

So far, we have discussed the classical theory of the gradient flow. In order to proceed to the quantum theory of the flowed gauge field, we consider the functional integral over the initial gauge field $A_\mu(x)$ as follows:

$$\begin{aligned}&\langle B_{\mu_1}(t_1,x_1)\cdots B_{\mu_n}(t_n,x_n)\rangle \\ =&\frac{1}{\mathcal{Z}}\int \mathcal{D}A_\mu \mathcal{D}c\mathcal{D}\bar{c}\, e^{-S_{\mathrm{G}}-S_{\mathrm{gf}}-S_{c\bar{c}}} B_{\mu_1}(t_1,x_1)\cdots B_{\mu_n}(t_n,x_n).\end{aligned} \tag{4.61}$$

For instance, when we consider the contraction $\langle B_\mu(t,x)B_\nu(s,y)\rangle$ to the lowest order, it corresponds to the free propagator of the flowed gauge field. We have to perform the following integration,

$$\begin{aligned}&\langle B_\mu(t,x)B_\nu(s,y)\rangle_0 \\ =&\int d^D z \int_p \frac{ip(x-z)}{p^2}[(\delta_{\mu\rho}p^2 - p_\mu p_\rho)e^{-tp^2} + p_\mu p_\rho e^{-\alpha_0 t p^2}] \\ &\times \int d^D w \int_q \frac{iq(y-z)}{q^2}[(\delta_{\nu\sigma}q^2 - q_\nu q_\sigma)e^{-sq^2} + q_\nu q_\sigma e^{-\alpha_0 s q^2}] \\ &\times g_0^2\delta^{ab}\int_k \frac{ik(z-w)}{(k^2)^2}\left[(\delta_{\rho\sigma}k^2 - k_\rho k_\sigma) + \frac{1}{\lambda_0}k_\rho k_\sigma\right].\end{aligned} \tag{4.62}$$

After simple integrations over z, w, and $p = -q = k$, we obtain the following propagator:

$$\langle B_\mu(t,x)B_\nu(s,y)\rangle_0$$
$$=g_0^2\delta^{ab}\int_p \frac{ip(x-y)}{(p^2)^2}\left[(\delta_{\mu\nu}p^2 - p_\mu p_\nu)e^{-(t+s)p^2} + \frac{1}{\lambda_0}p_\mu p_\nu e^{-\alpha_0(t+s)p^2}\right]. \quad (4.63)$$

We notice that the free propagator, Eq. (4.63) contains the Gaussian damping factors $e^{-(t+s)p^2}$ and $e^{-\alpha_0(t+s)p^2}$ stemmed from the heat kernel, which is a key for the UV finiteness of the correlation functions as discussed later.

4.2.3 *Example: Two-Point Function of Flowed Gauge Field*

Now we investigate the two-point function of the flowed gauge field $\langle B_\mu^a(t,x)B_\nu^b(s,y)\rangle$ to the one-loop order. We have already obtained the tree-level propagator as given in Eq. (4.63). The one-loop diagrams contributing to this calculation are summarized in Fig. 4.5. Here the black-filled circles and single wavy lines represent the YM vertices and the ordinary propagators in the D-dimensional space, respectively, although we drew the gauge propagator as stringy shape before. In Fig. 4.5, the diagrams (a)–(d) consist only of the ordinary YM vertices while the others have the flow vertices, which do not appear in D-dimensional YM theory discussed in the previous section.

First we consider Fig. 4.5(a)–(d). As calculated in the previous section, the diagram Fig. 4.5(c) vanishes and Fig. 4.5(d), the counter term arises from the parameter renormalizations,

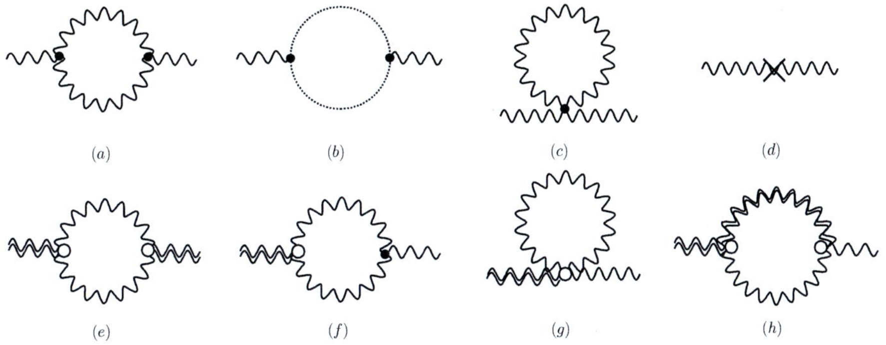

Fig. 4.5 Flow diagrams, which are needed to calculate the two-point function $\langle B_\mu^a(t,x)B_\nu^b(s,y)\rangle$ up to one-loop order

$$g_0^2 = \mu^{2\epsilon} g^2 \left[1 + \frac{g^2}{(4\pi)^2} \frac{1}{\epsilon} \left(-\frac{11}{3} \right) C_A \right], \tag{4.64}$$

$$\lambda_0 = \lambda \left[1 + \frac{g^2}{(4\pi)^2} \frac{1}{\epsilon} \left(-\frac{5}{3} \right) C_A \right]. \tag{4.65}$$

The only difference in the calculation is the additional Gaussian damping factor to the external lines. In what follows, we employ the Feynman gauge $\lambda_0 = \alpha_0 = 1$. In total, from the fact that α_0 does not receive the renormalization, we have

$$\begin{aligned} &\langle B_\mu^a(t,x) B_\nu^b(s,y) \rangle_{\text{1-loop, div.} + \text{parameter ren.}} \\ =&\mu^{2\epsilon} g^2 \delta^{ab} \int_p \frac{e^{ip(x-y)}}{(p^2)^2} e^{-(t+s)p^2} \frac{g^2}{(4\pi)^2} \frac{1}{\epsilon} (-2) C_A \delta_{\mu\nu} p^2. \end{aligned} \tag{4.66}$$

Note that the divergent term proportional to $1/\epsilon$ still remains because we have only renormalized the parameters g and λ. In the usual renormalization of the YM theory, we further include the wave function renormalization for the gauge field, so that the remaining divergence be absent.

In the present flowed system, however, we also have diagrams, Fig. 4.5(e)–(h), which contain the flow vertices instead of the wave function renormalization. As an example of the calculation of the flow diagrams, we evaluate Fig. 4.5(g). According to the Feynman rule, we can write down the diagram (g) and calculate it as follows:

$$\begin{aligned} &\int d^D z \int_0^t du \int_p e^{ip(x-z)} e^{-(t-u)p^2} f^{acd} f^{def} \langle B_\rho^c(u,z) B_\rho^e(u,z) B_\mu^f(u,z) B_\nu^b(s,y) \rangle_0 \\ =&g_0^4 \int d^D z \int_0^t du \int_p e^{ip(x-z)} e^{-(t-s)p^2} \int_q \frac{e^{iq(y-z)}}{q^2} e^{-(s+u)q^2} \int_l \frac{1}{l^2} e^{-2ul^2} \\ &\times (f^{acd} f^{dcb} D + f^{acd} f^{dbc}) \delta_{\mu\nu} \\ =&g_0^2 \delta^{ab} \int_p \frac{e^{ip(x-z)}}{p^2} e^{-(t+s)p^2} \int_0^t du \int_l \frac{1}{l^2} e^{-2ul^2} (-D+1) C_A \delta_{\mu\nu}. \end{aligned} \tag{4.67}$$

Thus, the flow time integral produces a singularity proportional to $1/\epsilon$,[3]

[3] In the dimensional regularization, we use the following fact:

$$\int_p \frac{1}{p^2} e^{-2up^2} = \frac{1}{(4\pi)^{D/2}} \frac{1}{D/2-1} (2u)^{-D/2+1},$$

$$\int_0^t du\, u^{-D/2+1} = \frac{1}{-D/2+2} t^{-D/2+2} \sim \frac{1}{\epsilon} + \ln t.$$

$$
\begin{aligned}
&(\text{Fig. } 4.5\,(g))|_{\text{1-loop, div}} \\
&= g_0^2 \delta^{ab} \int_p \frac{e^{ip(x-z)}}{p^2} e^{-(t+s)p^2} \left[\frac{g_0^2}{(4\pi)^2} \left\{ \frac{1}{\epsilon} + \frac{1}{2} \ln(4\pi t) + \frac{1}{2} \ln(4\pi s) \right\} (-3) C_A \delta_{\mu\nu} \right],
\end{aligned}
\tag{4.68}
$$

where included is the contribution of the diagram obtained by the exchange $(t, x, a) \leftrightarrow (s, y, b)$.

From the above expression, it is understood that, for a fixed flow time, the loop integrals are absolutely convergent due to the Gaussian damping factor. After the integration over the flow time, however, the Gaussian factors reduce to unity, so that the integrals generally exhibit UV divergences.

We can similarly evaluate the other diagrams by somewhat tedious calculations. After the calculation, in fact, the diagram Fig. 4.5(e) turns to be finite. Eventually, the sum of the flow line diagrams, Fig. 4.5(e)–(h) shows the singularity as follows:

$$
g_0^2 \delta^{ab} \int_p \frac{e^{ip(x-z)}}{p^2} e^{-(t+s)p^2} \left[\frac{g_0^2}{(4\pi)^2} \left\{ \frac{1}{\epsilon} + \frac{1}{2} \ln(4\pi t) + \frac{1}{2} \ln(4\pi s) \right\} \cdot 2 C_A \delta_{\mu\nu} \right].
\tag{4.69}
$$

The result shows that there occurs a precise cancellation of the divergences from the diagrams containing the YM vertices although the gauge field $B^a_\mu(t, x)$ remains 'bare.' This fact means that the flow plays, in some sense, a role of the wave function renormalization. Thus, we obtain the finite two-point function $\langle B^a_\mu(t, x) B^b_\nu(s, y) \rangle$ up to the one-loop order as far as $t, s > 0$:

$$
\begin{aligned}
&\langle B^a_\mu(t, x) B^b_\nu(s, y) \rangle \\
&= g^2 \delta^{ab} \int_p \frac{e^{ip(x-y)}}{(p^2)^2} e^{-(t+s)p^2} \left[1 + \frac{g^2}{(4\pi)^2} C_A \Big(\ln(4\pi \mu^2 t) + \ln(4\pi \mu^2 s) \Big) \right] \delta_{\mu\nu} p^2 + (\text{finite}).
\end{aligned}
\tag{4.70}
$$

Here are some remarks: (i) We need no wave function renormalizations once g and λ are appropriately renormalized. This is a noticeable feature because the flowed gauge field is given by some combination of the bare gauge field in the four-dimensional Euclidean space through the flow equation. (ii) After the parameter renormalization, the limit $t = s \to 0$ can be non-trivial, which means that the correlation function $\langle B^a_\mu(t, x) B^b_\nu(s, y) \rangle$ does not simply reduce to $\langle A^a_\mu(x) A^b_\nu(y) \rangle$. While the latter has UV divergence, the former is UV finite. Thus, the flow procedure and the renormalization do not commute with each other. (iii) When we extend our argument to include the fermion fields, it is found that the fermion field actually requires the wave function renormalization even in the gradient flow although we do not handle the fermion field in this dissertation.

In the next section, more generally, we see that any correlation functions of the flowed gauge fields are UV finite even at the same space point without the wave function renormalization once g and λ are renormalized. This fact indicates that they do not depend on the manner of the regularization. This is a key to obtain the universal representation of EMT as discussed later.

4.3 Proof of Renormalizability of Gradient Flow

In this section, we prove that any composite operators constructed from the flowed gauge fields $B^a_\mu(t,x)$ are automatically UV finite quantities. The proof of the renormalizability is based on the mathematical induction. We make use of the Ward–Takahashi (WT) identity derived from the BRS invariance of the vacuum represented in terms of the effective action, which is the generating functional for 1PI diagrams, hence that for the correlation functions. Then, we expand the effective action order by order with respect to the loop diagram, or the $\hbar$ expansion. We prove that the effective action at the $(l+1)$th order be finite, assuming that all of them up to the lth order are chosen to be finite with the appropriate renormalization factors Z, Z_3 and $\tilde{Z}_3$.

4.3.1 *$(D+1)$-Dimensional Field Theory*

For the proof, we construct the $(D+1)$-dimensional field theory, which reproduces the flow equation. With the introduction of the Lagrange multiplier $L_\mu(t,x) = L^a_\mu(t,x)T^a$, over which the integration imposes the flow equation, and the bulk ghost field $d(t,x)$, we consider the total action,

$$S = S_{\mathrm{G}} + S_{\mathrm{gf}} + S_{c\bar{c}} + S_{\mathrm{fl}} + S_{d\bar{d}}, \tag{4.71}$$

where the $(D+1)$-dimensional sections are given by

$$S_{\mathrm{fl}} = -2\int_0^\infty dt \int d^D x\, \mathrm{Tr}[L_\mu(t,x)(\partial_t B_\mu(t,x) - D_\nu G_{\nu\mu}(t,x) - \alpha_0 D_\mu \partial_\nu B_\nu(t,x))], \tag{4.72}$$

$$S_{d\bar{d}} = -2\int_0^\infty dt \int d^D x\, \mathrm{Tr}[\bar{d}(t,x)(\partial_t d(t,x) - \alpha_0 D_\mu \partial_\mu d(t,x)]. \tag{4.73}$$

The boundary conditions are

$$B_\mu(t=0,x) = A_\mu(x), \quad d(t=0,x) = c(x), \tag{4.74}$$

while $L_\mu(t=0,x)$ and $\bar{d}(t=0,x)$ are integrated freely as independent variables in the functional integral. Now we have to make the meaning of the integral and the derivative with respect to the flow time precise as follows:

$$\int_0^\infty dt \to \epsilon \sum_{t=0}^{\infty}, \quad \partial_t B_\mu(t,x) \to \frac{1}{\epsilon}\Big(B_\mu(t+\epsilon,x) - B_\mu(t,x)\Big), \tag{4.75}$$

which gives us the vanishing correlation at the same flow time. In particular, we have

$$\partial_t B_\mu(t=0,x) = \frac{1}{\epsilon}\Big(B_\mu(t=\epsilon,x) - A_\mu(x)\Big), \quad \partial_t d(t=0,x) = \frac{1}{\epsilon}\Big(d(t=\epsilon,x) - c(x)\Big). \tag{4.76}$$

The heat kernel is also given with the discretized flow time:

$$K_t^\epsilon(x)_{\mu\nu} = \int_p \frac{e^{ipx}}{p^2}[(\delta_{\mu\nu}p^2 - p_\mu p_\nu)(1-\epsilon p^2)^{t/\epsilon} + p_\mu p_\nu (1-\epsilon\alpha_0 p^2)^{t/\epsilon}], \tag{4.77}$$

which fulfills

$$\frac{1}{\epsilon}(K_{t+\epsilon}^\epsilon(x)_{\mu\nu} - K_t^\epsilon(x)_{\mu\nu}) = \partial^2 K_t^\epsilon(x)_{\mu\nu} + (\alpha_0 - 1)\partial_\mu\partial_\rho K_t^\epsilon(x)_{\rho\nu}, \tag{4.78}$$

with especially $K_0^\epsilon = \delta_{\mu\nu}\delta^D(x)$.

It is straightforward to derive the propagators from the action. The non-vanishing propagators are given by

$$\langle B_\mu^a(t,x) L_\nu^b(s,y)\rangle_0 = \delta^{ab}\theta(t-s)K_{t-s-\epsilon}^\epsilon(x-y)_{\mu\nu}, \tag{4.79}$$

$$\begin{aligned}&\langle B_\mu^a(t,x) B_\nu^b(s,y)\rangle_0 \\ =&g_0^2\delta^{ab}\int_p \frac{e^{ip(x-y)}}{(p^2)^2}\left[(\delta_{\mu\nu}p^2 - p_\mu p_\nu)(1-\epsilon p^2)^{(t+s)/\epsilon} + \frac{1}{\lambda_0}p_\mu p_\nu(1-\epsilon\alpha_0 p^2)^{(t+s)/\epsilon}\right],\end{aligned} \tag{4.80}$$

where $\theta(t)$ denotes the regularized step function such that $\theta(t) = 1$ for $t > 0$ and $\theta(t) = 0$ for $t \leq 0$. In the present prescription, the former propagator becomes the heat kernel or the flow line in the $\epsilon \to 0$ limit while the latter reproduces the ordinary $\langle B_\mu^a(t,x)B_\nu^b(s,y)\rangle_0$. As is obvious from the above expression, $\langle B_\mu^a(t,x)L_\nu^b(s,y)\rangle_0 = 0$ for $t \leq s$, which indicates that there are no loop diagrams containing only the heat kernels because any loops must consist of at least one flow loop satisfying $t \leq s$. Similarly, for the bulk ghost field, the two-point function is written by

$$\langle d^a(t,x)\bar{d}^b(s,y)\rangle_0 = \delta^{ab}\theta(t-s)K_{t-s-\epsilon}^\epsilon(x-y;\alpha_0), \tag{4.81}$$

where

$$K_t^\epsilon(x;\alpha_0) = \int_p e^{ipx}(1-\epsilon\alpha_0 p^2)^{t/\epsilon}. \tag{4.82}$$

In the $\epsilon \to 0$ limit, this results in the ordinary ghost propagator. In the same sense as mentioned above, we have $\langle d^a(t,x)\bar{d}^b(s,y)\rangle_0 = 0$ for $t \le 0$.

4.3.2 BRS Invariance and WT Identity of $(D+1)$-Dimensional System

The $(D+1)$-dimensional BRS transformation is defined by

$$\delta B_\mu(t,x) = D_\mu d(t,x), \tag{4.83}$$

$$\delta d(t,x) = -d^2(t,x), \tag{4.84}$$

$$\delta \bar{d}(t,x) = D_\mu L_\mu(t,x) - \{d,\bar{d}\}(t,x), \tag{4.85}$$

$$\delta L_\mu(t,x) = [L_\mu, d](t,x), \tag{4.86}$$

where δ is nilpotent as in the ordinary YM theory. Although the BRS transformation for $\bar{d}(t,x)$ seems to be curious, this field is dropped in the sense that it is regarded as the $(D+1)$-dimensional NL field. We can directly confirm that the BRS invariance of $S_{\text{fl}} + S_{d\bar{d}}$ in the $(D+1)$-dimensional system. Although the BRS symmetry is, in fact, broken by $O(\epsilon)$, it turns out that those breaking terms are harmless [21].

Now let us derive the WT identity. We introduce the source terms with respect to each field:

$$\begin{aligned} S_J = 2\int d^Dx\, \mathrm{Tr}[J^A_\mu(x)A_\mu(x) + J^c(x)c(x) + J^{\bar{c}}(x)\bar{c}(x) + J^B(x)B(x)] \\ +2 ⨍_0^\infty dt \int d^Dx\, \mathrm{Tr}[J^B_\mu(t,x)B_\mu(t,x) + J^d(t,x)d(t,x)] \\ +2\int_0^\infty dt \int d^Dx\, \mathrm{Tr}[J^L_\mu(t,x)L_\mu(t,x) + J^{\bar{d}}(t,x)\bar{d}(t,x)], \end{aligned} \tag{4.87}$$

where we define the ‘cut integral’:

$$⨍_0^\infty dt \equiv \epsilon \sum_{t=\epsilon}^{\infty}. \tag{4.88}$$

In particular, we prepare both the ordinary field and the flowed one for the gauge fields and the ghost fields. We also introduce the other source terms for the BRS transformation of each field:

$$
\begin{aligned}
S_K = 2 \int d^D x \, \mathrm{Tr}[K_\mu^A(x) D_\mu c(x) - K^c(x) c^2(x)] \\
+ 2 \int_0^\infty dt \int d^D x \, \mathrm{Tr}[K_\mu^B(t,x) D_\mu d(t,x) - K^d(t,x) d^2(t,x)] \\
+ 2 \int_0^\infty dt \int d^D x \, \mathrm{Tr}\Big[K_\mu^L(t,x)[L_\mu, d](t,x) \\
+ K^{\bar{d}}(t,x)\big(D_\mu L_\mu(t,x) - \{d, \bar{d}\}(t,x)\big)\Big]. \quad (4.89)
\end{aligned}
$$

Note that $\delta S_K = 0$ due to the nilpotency of δ. Totally, with the total action S, or Eq. (4.71), we consider $e^{-S-S_J-S_K}$. In terms of the connected correlation function $W[J, K]$, whose Legendre transformation gives the effective action, when we consider

$$
e^{-W[J,K]} = \int \mathcal{D}\mu \, e^{-S-S_J-S_K}, \quad (4.90)
$$

where $\mathcal{D}\mu$ represents the integral measure with respect to each field, it leads us to the following WT identity under the BRS transformation:

$$
\begin{aligned}
0 = &\Big\langle -2 \int d^D x \, \mathrm{Tr}[J_\mu^A(x) D_\mu c(x) + J^c(x) c^2(x) - J^{\bar{c}}(x) B(x)] \Big\rangle \\
&+ \Big\langle -2 \int_0^\infty dt \int d^D x \, \mathrm{Tr}[J_\mu^B(t,x) D_\mu d(t,x) + J^d(t,x) d^2(t,x)] \Big\rangle \\
&+ \Big\langle -2 \int_0^\infty dt \int d^D x \, \mathrm{Tr}\Big[J_\mu^L(t,x)[L_\mu, d](t,x) - J^{\bar{d}}\big(D_\mu L_\mu(t,x) - \{d, \bar{d}\}(t,x)\big)\Big] \Big\rangle.
\end{aligned} \quad (4.91)
$$

Here we use the fact that the total action S and the action S_K are invariant under δ while S_J is not. The above WT identity results from the BRS invariance of vacuum. In fact, $O(\epsilon)$ terms break the above WT identity. They, however, vanish in the $\epsilon \to 0$ limit, so that we neglect them by assuming the limit is taken in the appropriate stage [21].

Now we can evaluate the effective action Γ, the generating functional of 1PI diagrams in the presence of K. We perform the Legendre transformation with respect to only J, and not K:

$$
\Gamma[\phi, K] = W[J, K] + J \cdot \phi; \quad \frac{\delta}{\delta J} W[J, K] = -\langle \varphi \rangle = \phi, \quad (4.92)
$$

where we symbolically denote the fields as φ in

$$
e^{-W[J,K]} = \int d\varphi e^{-S + J^\varphi \cdot \varphi + K^\varphi \cdot \varphi}. \quad (4.93)
$$

In terms of the effective action, we have

$$\frac{\delta}{\delta\phi}\Gamma[\phi, K] = \pm J, \quad \frac{\delta}{\delta K}\Gamma[\phi, K] = \langle\delta\varphi\rangle. \tag{4.94}$$

Here the signs depend on whether ϕ is Grassmann even or odd. We have explicit formula for the effective action as

$$\begin{aligned}
&\Gamma[A_\mu, c, \bar{c}, B, K_\mu^A, K^c; B_\mu, d, L_\mu, \bar{d}, K_\mu^B, K^d, K_\mu^L, K^{\bar{d}}] \\
=&W[J, K] \\
&+2\int d^Dx\, \mathrm{Tr}[J_\mu^A(x)A_\mu(x) + J^c(x)c(x) + J^{\bar{c}}(x)\bar{c}(x) + J^B(x)B(x)] \\
&+2\int_0^\infty dt \int d^Dx\, \mathrm{Tr}[J_\mu^B(t,x)B_\mu(t,x) + J^d(t,x)d(t,x)] \\
&+2\int_0^\infty dt \int d^Dx\, \mathrm{Tr}[J_\mu^L(t,x)L_\mu(t,x) + J^{\bar{d}}(t,x)\bar{d}(t,x)].
\end{aligned} \tag{4.95}$$

Here and in the following, in the effective action Γ, the variables of the fields indicate not dynamical ones but the expectation values such as $\langle A_\mu(x)\rangle$ unless otherwise mentioned. Thus, the WT identity, Eq. (4.91), is rewritten in terms of Γ:

$$\begin{aligned}
&\int d^Dx \left[\frac{\delta\Gamma}{\delta A_\mu^a(x)}\frac{\delta\Gamma}{\delta K_\mu^{Aa}(x)} + \frac{\delta\Gamma}{\delta c^a(x)}\frac{\delta\Gamma}{\delta K^{ca}(x)} - \frac{\delta\Gamma}{\delta\bar{c}^a(x)}B^a(x)\right] \\
&+\int_0^\infty dt \int d^Dx \left[\frac{\delta\Gamma}{\delta B_\mu^a(t,x)}\frac{\delta\Gamma}{\delta K_\mu^{Ba}(t,x)} + \frac{\delta\Gamma}{\delta d^a(t,x)}\frac{\delta\Gamma}{\delta K^{da}(t,x)}\right] \\
&+\int_0^\infty dt \int d^Dx \left[\frac{\delta\Gamma}{\delta L_\mu^a(t,x)}\frac{\delta\Gamma}{\delta K_\mu^{La}(t,x)} + \frac{\delta\Gamma}{\delta\bar{d}^a(t,x)}\frac{\delta\Gamma}{\delta K^{\bar{d}a}(t,x)}\right] = 0.
\end{aligned} \tag{4.96}$$

We have, in the D-dimensional sense, the equation of motion or the Schwinger–Dyson equation,

$$\left\langle\frac{1}{g_0^2}\left[\partial_\mu A_\mu^a(x) - \frac{1}{\lambda_0}B^a(x)\right]\right\rangle - J^{Ba}(x) = 0, \tag{4.97}$$

$$\left\langle-\frac{1}{g_0^2}\partial_\mu D_\mu c^a(x)\right\rangle + J^{\bar{c}a}(x) = 0, \tag{4.98}$$

which is also expressed in terms of effective action,

$$\frac{\delta\Gamma}{\delta B^a(x)} = \frac{1}{g_0^2}\left[\partial_\mu A_\mu^a(x) - \frac{1}{\lambda_0}B^a(x)\right], \tag{4.99}$$

$$\frac{\delta\Gamma}{\delta\bar{c}^a(x)} - \frac{1}{g_0^2}\partial_\mu\frac{\delta\Gamma}{\delta K_\mu^{Aa}(x)} = 0. \tag{4.100}$$

By defining

$$\tilde{K}_\mu^{Aa}(x) = K_\mu^{Aa}(x) - \frac{1}{g_0^2}\partial_\mu \bar{c}^a(x), \tag{4.101}$$

the above relations show that the dependence of $B^a(x)$ and $\bar{c}^a(x)$ are completely eliminated:

$$\begin{aligned}&\tilde{\Gamma}[A_\mu, c, \tilde{K}_\mu^A, K^c; B_\mu, d, L_\mu, \bar{d}, K_\mu^B, K^d, K_\mu^L, K^{\bar{d}}]\\ =&\Gamma + \frac{2}{g_0^2}\int d^D x \operatorname{Tr}\left[B(x)\left(\partial_\mu A_\mu(x) - \frac{1}{2\lambda_0}B(x)\right)\right].\end{aligned} \tag{4.102}$$

Therefore, it is understood that the reduced effective action $\tilde{\Gamma}$ obeys the following WT identity:

$$\begin{aligned}&\int d^D x\left[\frac{\delta\tilde{\Gamma}}{\delta A_\mu^a(x)}\frac{\delta\tilde{\Gamma}}{\delta \tilde{K}_\mu^{Aa}(x)} + \frac{\delta\tilde{\Gamma}}{\delta c^a(x)}\frac{\delta\tilde{\Gamma}}{\delta K^{ca}(x)}\right]\\ &+\int_0^\infty dt\int d^D x\left[\frac{\delta\tilde{\Gamma}}{\delta B_\mu^a(t,x)}\frac{\delta\tilde{\Gamma}}{\delta K_\mu^{Ba}(t,x)} + \frac{\delta\tilde{\Gamma}}{\delta d^a(t,x)}\frac{\delta\tilde{\Gamma}}{\delta K^{da}(t,x)}\right]\\ &+\int_0^\infty dt\int d^D x\left[\frac{\delta\tilde{\Gamma}}{\delta L_\mu^a(t,x)}\frac{\delta\tilde{\Gamma}}{\delta K_\mu^{La}(t,x)} + \frac{\delta\tilde{\Gamma}}{\delta \bar{d}^a(t,x)}\frac{\delta\tilde{\Gamma}}{\delta K^{\bar{d}a}(t,x)}\right] = 0.\end{aligned} \tag{4.103}$$

4.3.3 *Proof of Renormalizability*

Now we come to the stage for the proof of the renormalizability of the flowed system with the use of the $(D+1)$-dimensional field theory [3, 21]. What we have to do is to prove, by induction, that the reduced effective action $\tilde{\Gamma}$ in terms of the renormalized quantities be finite once we appropriately choose Z, Z_3 and $\tilde{Z}_3$ order by order. The renormalized quantities are summarized below:

$$g_0^2 = \mu^{2\epsilon} g^2 Z, \tag{4.104}$$

$$A_\mu^a(x) = Z^{1/2}Z_3^{1/2}(A_\mathrm{R})_\mu^a(x), \quad \tilde{K}_\mu^{Aa}(x) = Z^{-1/2}Z_3^{-1/2}(\tilde{K}_\mathrm{R}^A)_\mu^a, \tag{4.105}$$

$$c^a(x) = \tilde{Z}_3 Z^{1/2}Z_3^{1/2}c_\mathrm{R}^a(x), \quad K^{ca}(x) = \tilde{Z}_3^{-1}Z^{-1/2}Z_3^{-1/2}K_\mathrm{R}^{ca}(x), \tag{4.106}$$

$$\lambda_0 = Z_3^{-1}\lambda, \tag{4.107}$$

$$B^a(x) = Z^{1/2}Z_3^{-1/2}B_\mathrm{R}^a, \tag{4.108}$$

$$\bar{c}^a(x) = Z^{1/2}Z_3^{-1/2}\bar{c}_\mathrm{R}^a(x). \tag{4.109}$$

Note again that the flowed gauge field $B_\mu^a(t,x)$ does not receive the wave function renormalization, hence the source term $K_\mu^{B\,a}(t,x)$ does not. In terms of the renormalized quantities, the WT identity is written by

$$\int d^Dx \left[\frac{\delta\tilde{\Gamma}}{\delta(A_{\mathrm{R}})_\mu^a(x)} \frac{\delta\tilde{\Gamma}}{\delta(\tilde{K}_{\mathrm{R}}^A)_\mu^a(x)} + \frac{\delta\tilde{\Gamma}}{\delta c_{\mathrm{R}}^a(x)} \frac{\delta\tilde{\Gamma}}{\delta K_{\mathrm{R}}^{c\,a}(x)} \right]$$
$$+ \int_0^\infty dt \int d^Dx \left[\frac{\delta\tilde{\Gamma}}{\delta B_\mu^a(t,x)} \frac{\delta\tilde{\Gamma}}{\delta K_\mu^{B\,a}(t,x)} + \frac{\delta\tilde{\Gamma}}{\delta d^a(t,x)} \frac{\delta\tilde{\Gamma}}{\delta K^{d\,a}(t,x)} \right]$$
$$+ \int_0^\infty dt \int d^Dx \left[\frac{\delta\tilde{\Gamma}}{\delta L_\mu^a(t,x)} \frac{\delta\tilde{\Gamma}}{\delta K_\mu^{L\,a}(t,x)} + \frac{\delta\tilde{\Gamma}}{\delta \bar{d}^a(t,x)} \frac{\delta\tilde{\Gamma}}{\delta K^{\bar{d}\,a}(t,x)} \right] = 0. \tag{4.110}$$

For the proof, we consider the loop expansion, or the $\hbar$ expansion of $\tilde{\Gamma}$ in the renormalized perturbation theory:

$$\tilde{\Gamma} = \sum_{l=0}^\infty \tilde{\Gamma}^{(l)}. \tag{4.111}$$

In particular, the tree-level effective action $\tilde{\Gamma}^{(0)}$ is given by

$$\tilde{\Gamma}^{(0)} = (S + S_{\mathrm{fl}} + S_{d\bar{d}} + S_K)|_{Z=Z_3=\tilde{Z}_3=1,\, K_\mu^A \to (\tilde{K}_{\mathrm{R}}^A)_\mu}, \tag{4.112}$$

since both $\bar{c}^a(x)$ and $B^a(x)$ are eliminated and the Legendre transformation is completed with respect to J, hence the absence of S_{gf}, $S_{c\bar{c}}$, and S_J. In addition, we have also the counter term ΔS in the renormalized perturbation theory,

$$\Delta S = S + S_{\mathrm{fl}} + S_{d\bar{d}} + S_K - (S + S_{\mathrm{fl}} + S_{d\bar{d}} + S_K)|_{Z=Z_3=\tilde{Z}_3=1,\, K_\mu^A \to (\tilde{K}_{\mathrm{R}}^A)_\mu}, \tag{4.113}$$

which contains the boundary counter term, that is the one in the D-dimensional system,

$$\Delta S_{\mathrm{b.c.}} = 2 \int d^Dx\, \mathrm{Tr}[L_\mu(0,x)(Z^{1/2} Z_3^{1/2} - 1)(A_{\mathrm{R}})_\mu(x)]$$
$$+ 2 \int d^Dx\, \mathrm{Tr}[\bar{d}(0,x)(\tilde{Z}_3 Z^{1/2} Z^1/2 - 1)c_{\mathrm{R}}(x)]. \tag{4.114}$$

We also note that $\tilde{\Gamma}^{(0)}$ correctly yields the tree-level propagators.

The proof of the renormalizability is carried out by induction, where it is argued that the divergent part of $\tilde{\Gamma}^{(l)}$ satisfies the WT relation, or the renormalization equation,

$$\tilde{\Gamma}^{(0)} * \tilde{\Gamma}^{(l+1),\,\text{div.}} = 0, \tag{4.115}$$

assuming that the Z, Z_3 and $\tilde{Z}_3$ up to the lth order are appropriately chosen so that $\tilde{\Gamma}^{(l)}$ is finite. Here we define the $*$ operator such that

$$\begin{aligned}
\tilde{\Gamma}^{(0)} * = &- \int d^D x \left[\frac{\delta \tilde{\Gamma}^{(0)}}{\delta (A_{\mathrm{R}})^a_\mu(x)} \frac{\delta}{\delta (\tilde{K}^A_{\mathrm{R}})^a_\mu(x)} + \frac{\delta \tilde{\Gamma}^{(0)}}{\delta (\tilde{K}^A_{\mathrm{R}})^a_\mu(x)} \frac{\delta}{\delta (A_{\mathrm{R}})^a_\mu(x)} \right. \\
&\left. + \frac{\delta \tilde{\Gamma}^{(0)}}{\delta c^a_{\mathrm{R}}(x)} \frac{\delta}{\delta K^{ca}_{\mathrm{R}}(x)} + \frac{\delta \tilde{\Gamma}^{(0)}}{\delta K^{ca}_{\mathrm{R}}(x)} \frac{\delta}{\delta c^a_{\mathrm{R}}(x)} \right] \\
&- \epsilon \int d^D x \left[\frac{\delta \tilde{\Gamma}^{(0)}}{\delta L^a_\mu(0,x)} \frac{\delta}{\delta K^{La}_\mu(0,x)} + \frac{\delta \tilde{\Gamma}^{(0)}}{\delta K^{La}_\mu(0,x)} \frac{\delta}{\delta L^a_\mu(0,x)} \right. \\
&\left. + \frac{\delta \tilde{\Gamma}^{(0)}}{\delta \bar{d}^a(0,x)} \frac{\delta}{\delta K^{\bar{d}a}(0,x)} + \frac{\delta \tilde{\Gamma}^{(0)}}{\delta K^{\bar{d}a}(0,x)} \frac{\delta}{\delta \bar{d}^a(0,x)} \right] \\
&+ (\text{other derivatives w.r.t the field variables at } t \neq 0).
\end{aligned} \tag{4.116}$$

In fact, the WT identity is simply denoted by

$$\tilde{\Gamma} * \tilde{\Gamma} = 0 \tag{4.117}$$

with the $*$ operator and, at the $(l+1)$th order,

$$\tilde{\Gamma}^{(0)} * \tilde{\Gamma}^{(l+1)} + \tilde{\Gamma}^{(1)} * \tilde{\Gamma}^{(l)} + \cdots = 0. \tag{4.118}$$

The assumption of the induction process lets the terms other than the first one be finite. Thus, it is sufficient to prove Eq. (4.115). Note that $\tilde{\Gamma}^{(0)}*$ is nilpotent due to the BRS invariance of $(S + S_{\text{fl}} + S_{d\bar{d}} + S_K)$.

We now decompose $\tilde{\Gamma}^{(0)}*$ into the operators in the D-dimensional theory and the $(D+1)$-dimensional one:

$$\tilde{\Gamma}^{(0)} * = \tilde{\Gamma}^{(0)}_D * + \tilde{\Gamma}^{(0)}_{D+1} *, \tag{4.119}$$

where the D-dimensional contribution is written by the ordinary operator,

$$\begin{aligned}
\tilde{\Gamma}^{(0)}_D * = &- \int d^D x \left[\frac{\delta \tilde{\Gamma}^{(0)}|_{L_\mu = \bar{d} = 0}}{\delta (A_{\mathrm{R}})^a_\mu(x)} \frac{\delta}{\delta (\tilde{K}^A_{\mathrm{R}})^a_\mu(x)} + \frac{\delta \tilde{\Gamma}^{(0)}|_{L_\mu = \bar{d} = 0}}{\delta (\tilde{K}^A_{\mathrm{R}})^a_\mu(x)} \frac{\delta}{\delta (A_{\mathrm{R}})^a_\mu(x)} \right. \\
&\left. + \frac{\delta \tilde{\Gamma}^{(0)}|_{L_\mu = \bar{d} = 0}}{\delta c^a_{\mathrm{R}}(x)} \frac{\delta}{\delta K^{ca}_{\mathrm{R}}(x)} + \frac{\delta \tilde{\Gamma}^{(0)}|_{L_\mu = \bar{d} = 0}}{\delta K^{ca}_{\mathrm{R}}(x)} \frac{\delta}{\delta c^a_{\mathrm{R}}(x)} \right].
\end{aligned} \tag{4.120}$$

We also decompose $\tilde{\Gamma}^{(l+1),\,\text{div.}}$ such that

$$\tilde{\Gamma}^{(l+1),\,\mathrm{div.}} = \tilde{\Gamma}_D^{(l+1),\,\mathrm{div.}} + \tilde{\Gamma}_{D+1}^{(l+1),\,\mathrm{div.}}, \tag{4.121}$$

where $\tilde{\Gamma}_D^{(l+1),\,\mathrm{div.}}$ does not contain the bulk fields L_μ and $\bar{d}$ while $\tilde{\Gamma}_{D+1}^{(l+1),\,\mathrm{div.}}$ do at least one bulk field. Then the renormalization equation (4.115) is decomposed into

$$\tilde{\Gamma}_D^{(0)} * \tilde{\Gamma}_D^{(l+1),\,\mathrm{div.}} = 0, \tag{4.122}$$

$$\tilde{\Gamma}_D^{(0)} * \tilde{\Gamma}_{D+1}^{(l+1),\,\mathrm{div.}} + \tilde{\Gamma}_{D+1}^{(0)} * \tilde{\Gamma}_D^{(l+1),\,\mathrm{div.}} + \tilde{\Gamma}_{D+1}^{(0)} * \tilde{\Gamma}_{D+1}^{(l+1),\,\mathrm{div.}} = 0, \tag{4.123}$$

where the former is the familiar renormalization equation in the D-dimensional field theory. The general solution of Eq. (4.122) is well known as [22]

$$\begin{aligned}\tilde{\Gamma}_D^{(l+1),\,\mathrm{div.}} = &-\frac{1}{2\mu^{2\epsilon}g^2}\int d^Dx\,\mathrm{Tr}[x_1(G_\mathrm{R})_{\mu\nu}(x)(G_\mathrm{R})_{\mu\nu}(x)]\\ &-\tilde{\Gamma}_D^{(0)} * \int d^Dx\,\mathrm{Tr}[y_1(\tilde{K}_\mathrm{R}^A)_\mu(x)(A_\mathrm{R})_\mu(x) + y_2K_\mathrm{R}^c(x)c_\mathrm{R}(x)],\end{aligned} \tag{4.124}$$

where x_1, y_1, and y_2 are divergent constant and

$$(G_\mathrm{R})_{\mu\nu}(x) = \partial_\mu(A_\mathrm{R})_\nu(x)\partial_\nu(A_\mathrm{R})_\mu(x) + [(A_\mathrm{R})_\mu,(A_\mathrm{R})_\nu](x). \tag{4.125}$$

Then, inserting Eq. (4.124) into Eq. (4.121), we obtain

$$\begin{aligned}\tilde{\Gamma}^{(l+1),\,\mathrm{div.}} =& -\frac{1}{2\mu^{2\epsilon}g^2}\int d^Dx\,\mathrm{Tr}[x_1(G_\mathrm{R})_{\mu\nu}(x)(G_\mathrm{R})_{\mu\nu}(x)]\\ &-2\tilde{\Gamma}_D^{(0)} * \int d^Dx\,\mathrm{Tr}[y_1(\tilde{K}_\mathrm{R}^A)_\mu(x)(A_\mathrm{R})_\mu(x) + y_2K_\mathrm{R}^c(x)c_\mathrm{R}(x)]\\ &+\tilde{\Gamma}_{D+1}^{(l+1),\,\mathrm{div.}} \qquad (4.126)\\ =& -\frac{1}{2\mu^{2\epsilon}g^2}\int d^Dx\,\mathrm{Tr}[x_1(G_\mathrm{R})_{\mu\nu}(x)(G_\mathrm{R})_{\mu\nu}(x)]\\ &-2\tilde{\Gamma}^{(0)} * \int d^Dx\,\mathrm{Tr}[y_1(\tilde{K}_\mathrm{R}^A)_\mu(x)(A_\mathrm{R})_\mu(x) + y_2K_\mathrm{R}^c(x)c_\mathrm{R}(x)]\\ &+\tilde{\tilde{\Gamma}}_{D+1}^{(l+1),\,\mathrm{div.}}, \qquad (4.127)\end{aligned}$$

where

$$\tilde{\tilde{\Gamma}}_{D+1}^{(l+1),\,\mathrm{div.}} = \tilde{\Gamma}_{D+1}^{(l+1),\,\mathrm{div.}} + 2\int d^Dx\,\mathrm{Tr}[y_1L_\mu(0,x)(A_\mathrm{R})_\mu(x) - y_2\bar{d}(0,x)c_\mathrm{R}(x)]. \tag{4.128}$$

The candidates of the divergent terms are encoded in the boundary integral, which must contain at least one L_μ or $\bar{d}$ with some constants z_1 and z_2, as follows:

$$\tilde{\tilde{\Gamma}}_{D+1}^{(l+1),\,\mathrm{div.}} = -2 \int d^D x \,\mathrm{Tr}[z_1 L_\mu(0,x)(A_\mathrm{R})_\mu(x) + z_2 \bar{d}(0,x) c_\mathrm{R}(x)]. \quad (4.129)$$

Then the WT identity, Eq. (4.115) yields

$$\begin{aligned}\tilde{\Gamma}^{(0)} * \tilde{\tilde{\Gamma}}_{D+1}^{(l+1),\,\mathrm{div.}} = &-2 \int d^D x \,\mathrm{Tr}[(z_1 - z_2) L_\mu(0,x) \partial_\mu c_\mathrm{R}(x) \\ &- z_2 L_\mu(0,x)[(A_\mathrm{R})_\mu, c_\mathrm{R}](x) - z_2 \bar{d}(t,x) c_\mathrm{R}^2(x)] = 0. \quad (4.130)\end{aligned}$$

From this relation, we deduce $z_1 = z_2 = 0$, so that $\tilde{\tilde{\Gamma}}_{D+1}^{(l+1),\,\mathrm{div.}} = 0$. Thus, the remaining divergence in $\tilde{\Gamma}^{(l+1),\,\mathrm{div.}}$ represents

$$\begin{aligned}\tilde{\Gamma}^{(l+1),\,\mathrm{div.}} = &-\frac{1}{2\mu^{2\epsilon} g^2} \int d^D x \,\mathrm{Tr}[x_1 (G_\mathrm{R})_{\mu\nu}(x)(G_\mathrm{R})_{\mu\nu}(x)] \\ &- 2\tilde{\Gamma}^{(0)} * \int d^D x \,\mathrm{Tr}[y_1 (\tilde{K}_\mathrm{R}^A)_\mu(x)(A_\mathrm{R})_\mu(x) + y_2 K_\mathrm{R}^c(x) c_\mathrm{R}(x)].\end{aligned} \quad (4.131)$$

One can confirm that the above possible divergence at $(l+1)$th order can be canceled by choosing in the $(l+1)$th order,

$$Z^{(l+1)} = x_1 \quad (4.132)$$

$$Z_3^{(l+1)} = -x_1 + 2y_1 \quad (4.133)$$

$$\tilde{Z}_3^{(l+1)} = -y_1 - y_2. \quad (4.134)$$

This is possible because the number of the divergent constants is three. We now come to the end of the proof, in which we conclude that we do need no wave function renormalization of the flowed gauge field after the others appropriately renormalized.

4.4 Application to Energy Momentum Tensor

In this section, we consider the universal formula of EMT on the basis of the gradient-flow method. Generally, the Noether's theorem insists that when a dynamical system is invariant under a continuous transformation, it has corresponding conserved quantities. In a field theory, there exists conserved currents. In particular, EMT is the conserved current under the translational symmetry with respect to the space and time in the quantum field theory.

Later, we investigate the spatial distribution of EMT around static source(s) by means of numerical simulations based on the lattice gauge theory, where we implement the space–time discretization, which gives rise to the regularization, so-called the lattice regularization. EMT is a key observable for us to investigate the local property of the quantum field. The lattice gauge theory and its simulations on the lattice regularization provide us opportunities to explore quantum systems non-perturbatively. We would like to construct EMT on the lattice, which automatically becomes the correct EMT in the continuum limit $a \to 0$. However, since the lattice regularization explicitly breaks such a continuous space–time symmetry, it is difficult to define and numerically measure EMT on the lattice.[4] In that case, the possibility comes to mind to make use of the dimensional regularization, which strictly preserve the translational symmetry as well as the gauge symmetry, so that we can write down the EMT formula. Yet again, for the non-perturbative analyses we would like, it is impossible to employ the EMT in the framework of the dimensional regularization because it is based on the perturbation.

In order to overcome the difficulty raised above, the gradient flow plays crucial roles. The idea is as follows: We construct the renormalized EMT with the use of the flowed fields, so that it is consistent with the EMT formula in the dimensional regularization. As proved in the previous section, such flowed composite operators are UV finite, hence the EMT formula obtained in this manner is independent of specific kinds of regularization. Therefore, it is applicable even to the lattice regularization because such quantities have the same expression in the limit $\epsilon \to 0$ in the dimensional regularization as well as $a \to 0$ in the lattice regularization. In this sense, the gradient flow plays a role to bridge the different regularizations. The remaining matter is how we construct such EMT operator with the combination of the flowed fields. The following is mainly dedicated to that point.

4.4.1 Small Flow Time Expansion

We consider the action S_{G}. In the dimensional regularization, which respect the translational symmetry of the system, the symmetric EMT is defined as

$$\mathcal{T}_{\mu\nu}(x) = \frac{1}{g_0^2}\left[G^a_{\mu\rho}(x)G^a_{\nu\rho}(x) - \frac{1}{4}\delta_{\mu\nu}G^a_{\rho\sigma}(x)G^a_{\rho\sigma}(x)\right]. \tag{4.135}$$

The renormalized EMT is given by subtracting its possibly divergent expectation values:

$$\mathcal{T}^{\mathrm{R}}_{\mu\nu}(x) = \mathcal{T}_{\mu\nu} - \langle \mathcal{T}_{\mu\nu}(x)\rangle. \tag{4.136}$$

[4] In fact, it is possible to construct EMT on the lattice by means of, e.g., the shifted boundary method [23–26]. In such a formalism, the renormalized EMT formula receives non-trivial corrections to Eq. (4.136), which result from and depend on the lattice action one uses and the manner of the discretization of operators [27, 28]. The coefficients, which are contained in the renormalized EMT formula, have been already determined even for the full QCD [26]. Thus, there are a lot of outcomes obtained in this framework such as thermodynamic quantities. In contrast to such direction, since we seek a universal formula of EMT independent of the regularization, we here consider to construct EMT on the basis of the gradient-flow method.

We now need to find a composite operator of the flowed gauge fields which reduces to the EMT in the dimensional regularization. In general, the relation between the composite operators in the bulk region, $t > 0$ and those in the D-dimensional space is not obvious at all. Nevertheless, the relation becomes under control when the flow time is small. This idea is called the small flow time expansion (SFtX) and similar to the familiar operator product expansion. The SFtX method argues that, at $t \sim 0$, the operator constructed in the $(D+1)$-dimensional space is represented by the asymptotic series of the renormalized operators in the D-dimensional space:

$$O_{i\mu\nu}(t,x) = \langle O_{i\mu\nu}(t,x)\rangle \mathbb{1} + \sum_j \zeta_{ij}(t)[O_{j\mu\nu}(x) - \langle O_{j\mu\nu}(x)\rangle \mathbb{1}] + O(t), \quad (4.137)$$

where $O_{i\mu\nu}(x)$ and $O_{i\mu\nu}(t,x)$ denote some observable distinguished by the subscript i in D and $D+1$ dimension, and ζ_{ij} is an expansion coefficient as a function of t. The $O(t)$ term represents the dimension $(D+2)$ and the higher operators. This relation results from the fact that $O_{i\mu\nu}(t,x)$ with $t \to 0$ is a local quantity in the D-dimensional space and, at the same time, it is automatically UV finite due to the smearing of the gauge field within the radius $\sim \sqrt{2t}$ as long as $t > 0$. Inverting Eq. (4.137), we obtain

$$O_{i\mu\nu}(x) - \langle O_{i\mu\nu}(x)\rangle \mathbb{1} = \lim_{t\to 0}\left[\sum_j (\zeta^{-1})_{ij}(t)(O_{j\mu\nu}(t,x) - \langle O_{j\mu\nu}(t,x)\rangle \mathbb{1})\right], \quad (4.138)$$

which mentions that the D-dimensional operator can be extracted in $t \to 0$ if we know the $t \to 0$ behavior of the expansion coefficients $\zeta_{ij}(t)$.

For the construction of EMT, we consider the D-dimensional operators. In the $(D+1)$-dimensional space, there are two kinds of gauge-invariant D-dimensional operators:

$$U_{\mu\nu}(t,x) = G^a_{\mu\rho}(t,x)G^a_{\nu\rho}(t,x) - \frac{1}{4}\delta_{\mu\nu}G^a_{\rho\sigma}(t,x)G^a_{\rho\sigma}(t,x) \quad (4.139)$$

$$E(t,x) = \frac{1}{4}G^a_{\mu\nu}(t,x)G^a_{\mu\nu}(t,x). \quad (4.140)$$

Since we know EMT as the renormalized D-dimensional operator, according to the SFtX statement, they are expressed as follows:

$$U_{\mu\nu}(t,x) = \alpha_U(t)\left[\mathcal{T}^{\mathrm{R}}_{\mu\nu}(x) - \frac{1}{4}\delta_{\mu\nu}\mathcal{T}^{\mathrm{R}}_{\rho\rho}(x)\right] + O(t) \quad (4.141)$$

$$E(t,x) = \langle E(t,x)\rangle + \alpha_E(t)\mathcal{T}^{\mathrm{R}}_{\rho\rho}(x) + O(t), \quad (4.142)$$

where $\alpha_U(t)$ and $\alpha_E(t)$ are the expansion coefficients for $U_{\mu\nu}(t,x)$ and $E(t,x)$, respectively. Therefore, we acquire the renormalized EMT formula via the composite operators represented by flowed gauge fields,

$$\mathcal{T}^{\mathrm{R}}_{\mu\nu}(x) = \lim_{t\to 0} \mathcal{T}_{\mu\nu}(t,x),$$
$$\mathcal{T}_{\mu\nu}(t,x) = \frac{1}{\alpha_U(t)} U_{\mu\nu}(t,x) + \frac{\delta_{\mu\nu}}{4\alpha_E(t)}(E(t,x) - \langle E(t,x)\rangle). \tag{4.143}$$

The remaining task is to determine the behaviors of $\alpha_U(t)$ and $\alpha_E(t)$ at $t \sim 0$.

4.4.2 Renormalization Group Equation and Expansion Coefficients

We are interested in the $t \to 0$ behavior of the coefficients $\zeta_{ij}(t)$ in Eq. (4.137), or specifically $\alpha_U(t)$ in Eq. (4.141) and $\alpha_E(t)$ in Eq. (4.142). When all the composite operators in Eq. (4.137) are constructed from bare quantities, the operation of $(\mu\partial_\mu)_0$ on the both sides of Eq. (4.137) lets us obtain

$$\left(\mu\frac{\partial}{\partial\mu}\right)_0 \zeta_{ij}(t) = \left(\mu\frac{\partial}{\partial\mu} + \beta\frac{\partial}{\partial g}\right)\zeta_{ij}(t) = 0, \tag{4.144}$$

where μ indicates the renormalization scale and '0' means the differentiation with the fixed bare parameter. Here we use the fact that the bare quantities and the renormalized EMT do not receive the renormalization. Equation (4.144) states that $\zeta_{ij}(t)$ is independent of μ when we employ the running coupling $\bar{g}(q)$, which is defined as

$$q\frac{d\bar{g}(q)}{dq} = \beta(\bar{g}(q)); \quad \bar{g}(q=\mu) = g. \tag{4.145}$$

We choose the renormalization scale $\mu \propto 1/\sqrt{t}$. In the $t \to 0$ limit, where the coupling becomes weak due to the asymptotic freedom, we can evaluate $\zeta_{ij}(t)$ perturbatively. Reference [4] gives the one-loop order coefficients and now the two-loop-order coefficients are available shown in Refs. [16, 29]. The latter enables us to perform the numerical analysis with smaller systematic errors [16]. We here omit the details in the calculation of the coefficients, which requires to write down a correct set of diagrams and tedious calculations. We comment on the choice of μ. Although the coefficients are, in principal, independent of the choice of μ in the infinite-order perturbation, there are appropriate values for the actual calculations based on the fixed-order perturbation theory, where the possibility of the perturbative uncertainty exists, depending on the choice. A conventional choice is $\mu = 1/\sqrt{8t}$, and $\mu = 1/\sqrt{2e^{\gamma_E}t}$ is also proposed. The latter can somewhat simplify the form of the coefficients [16, 29].

4.4.3 Energy Momentum Tensor on the Lattice

The EMT formula, Eq. (4.143) contains only UV finite quantities and the universal coefficients; hence, it is independent of the regularization. Therefore, we can apply Eq. (4.143) for the lattice gauge theory based on the lattice regularization, leading to the numerical calculations on the lattice. Notice that the above universality is held once the cutoff is removed, in this case $a \to 0$. Thus, we must pay attention to the order of the limit: $a \to 0$ followed by $t \to 0$. Although we will explain specific procedure of the double extrapolation, $a \to 0$ and $t \to 0$ with our actual data, we here mention it in general.

The suppression of the lattice discretization effect by the sufficient smearing of the gradient flow is represented as the following requirement

$$a/2 \ll \sqrt{2t}. \tag{4.146}$$

The red-shaded area on t-a^2 plane in Fig. 4.6 is prohibited by Eq. (4.146). On the other hand, the upper bound of $\sqrt{2t}$ may be limited by the lambda parameter $1/\Lambda$, the temporal periodicity $1/(2T)$, and so on, depending on systems under consideration as seen in later chapters.

Keeping the above criterions on the appropriate window, we employ the following Ansatz for EMT,

$$\langle \mathcal{T}_{\mu\nu}(t,x)\rangle^{\text{lat}} = \langle \mathcal{T}^{\text{R}}_{\mu\nu}(x)\rangle + b_{\mu\nu}(t)a^2 + \sum_n c^{(n)}_{\mu\nu} t^n, \tag{4.147}$$

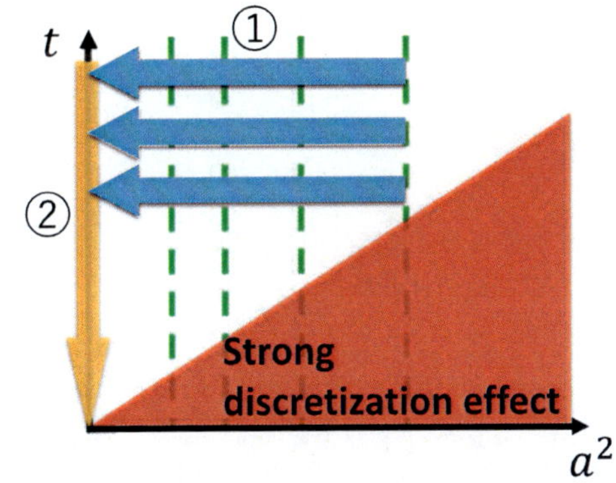

Fig. 4.6 Procedure of the double extrapolation on t-a^2 plane. We keep the condition $a/2 \ll \sqrt{2t}$ to suppress the discretization effect, which means to avoid data in the red-shaded area. According to Eq. (4.147), we first take $a \to 0$ limit (blue arrows) with fixed flow time (green dashed line). Second, we take $t \to 0$ limit (yellow arrow), so that we obtain the renormalized EMT in the ordinary Euclidean space

where the contributions from the discretization effect $b_{\mu\nu}$ comes from the plaquette while terms with respect to $c_{\mu\nu}^{(n)}$ from the SFtX method.[5] Thus, as explicitly shown in Fig. 4.6, after we take $a \to 0$ limit to remove the cutoff according to Eq. (4.147) with fixed t, then we perform the small t limit under Eq. (4.146) with appropriate windows.

4.5 Comments on Fermions

So far, we mainly focus on the gauge field and its smearing via the gradient flow. Here we briefly make some remarks on the fermion field.

As is similar to the case of the gauge field, the evolution of the fermion field along the flow time direction is feasible. In Ref. [30], the following choice of the fermion flow equation is suggested:

$$\partial_t \chi_f(t,x) = [D_\mu D_\mu - \alpha_0 \partial_\mu B_\mu(t,x)]\chi_f(t,x), \tag{4.148}$$

$$\partial_t \bar{\chi}_f(t,x) = \bar{\chi}_f(t,x)[\overleftarrow{D}_\mu \overleftarrow{D}_\mu - \alpha_0 \partial_\mu B_\mu(t,x)], \tag{4.149}$$

where the flowed fermion field is denoted by $\chi_f(t,x)$, whose flavor is f. The initial condition of the flowed fermion field is given by

$$\chi_f(t=0,x) = \psi_f(x), \quad \bar{\chi}_f(t=0,x) = \bar{\psi}_f(x), \tag{4.150}$$

with the ordinary fermion field $\psi_f(x)$ in D-dimensional Euclidean space. It turns out that the flowed fermion field requires the wave function renormalization as

$$\chi_{f\mathrm{R}}(t,x) = Z_{\chi_f}^{1/2}\chi_f(t,x), \quad \bar{\chi}_{f\mathrm{R}}(t,x) = Z_{\chi_f}^{1/2}\bar{\chi}_f(t,x), \tag{4.151}$$

with

$$Z_{\chi_f} = 1 + \frac{g^2}{(4\pi)^2} 3C_F \frac{1}{\epsilon}. \tag{4.152}$$

The argument on the renormalizability discussed above can be appropriately modified with the fermion field [5].

We finally consider the EMT defined via the gradient flow, including not only the gauge field but the fermion field. Under the dimensional regularization, the symmetric EMT is written down as

[5] Although $c_{\mu\nu}^{(n)}$ may in principle have logarithmic behavior in t, we omit that dependence in the following analyses.

$$\mathcal{T}_{\mu\nu}(x) = \frac{1}{g_0^2}\left[G^a_{\mu\rho}(x)G^a_{\nu\rho}(x) - \frac{1}{4}\delta_{\mu\nu}G^a_{\rho\sigma}(x)G^a_{\rho\sigma}(x)\right] + \frac{1}{4}\bar{\psi}(x)(\gamma_\mu \overleftrightarrow{D}_\nu + \gamma_\nu \overleftrightarrow{D}_\mu)\psi(x) - \frac{1}{2}\delta_{\mu\nu}\bar{\psi}\gamma_\rho \overleftrightarrow{D}_\rho\psi(x) - \delta_{\mu\nu}m_0\bar{\psi}(x)\psi(x), \tag{4.153}$$

with $\overleftrightarrow{D} \equiv D_\mu - \overleftarrow{D}_\mu$. The renormalized EMT is also defined by appropriately subtracting the vacuum expectation value. For the same reasoning as discussed above with much more tedious calculations on the matching coefficients, the SFtX leads to the correctly normalized EMT given by the following formula [5]:

$$\begin{aligned}
\mathcal{T}^{\mathrm{R}}_{\mu\nu}(x) = \lim_{t\to 0}\Big\{ & c_1(t)\left[O_{1\,\mu\nu}(t,x) - \frac{1}{4}O_{2\,\mu\nu}(t,x)\right] \\
& + c_2(t)[O_{2\,\mu\nu}(t,x) - \langle O_{2\,\mu\nu}(t,x)\rangle_0] \\
& + c_3(t)\sum_f [O^f_{3\,\mu\nu}(t,x) - 2O_{4\,\mu\nu}(t,x) - \langle O^f_{3\,\mu\nu}(t,x) - 2O_{4\,\mu\nu}(t,x)\rangle_0] \\
& + c_4(t)\sum_f [O^f_{4\,\mu\nu}(t,x) - \langle O^f_{4\,\mu\nu}(t,x)\rangle_0] \\
& + c_5(t)\sum_f [O^f_{5\,\mu\nu}(t,x) - \langle O^f_{5\,\mu\nu}(t,x)\rangle_0]\Big\},
\end{aligned} \tag{4.154}$$

with

$$O_{1\,\mu\nu}(t,x) = G^a_{\mu\rho}(t,x)G^a_{\nu\rho}(t,x), \tag{4.155}$$

$$O_{2\,\mu\nu}(t,x) = \delta_{\mu\nu}G^a_{\rho\sigma}(t,x)G^a_{\rho\sigma}(t,x), \tag{4.156}$$

$$O^f_{3\,\mu\nu}(t,x) = \mathring{\bar{\chi}}_f(t,x)(\gamma_\mu \overleftrightarrow{D}_\nu + \gamma_\nu \overleftrightarrow{D}_\mu)\mathring{\chi}_f(t,x), \tag{4.157}$$

$$O^f_{4\,\mu\nu}(t,x) = \delta_{\mu\nu}\mathring{\bar{\chi}}_f(t,x)\gamma_\mu \overleftrightarrow{D}_\mu\mathring{\chi}_f(t,x), \tag{4.158}$$

$$O^f_{5\,\mu\nu}(t,x) = \delta_{\mu\nu}m^f_0\mathring{\bar{\chi}}_f(t,x)\mathring{\chi}_f(t,x). \tag{4.159}$$

In order to avoid the complication on the wave function renormalization of the flowed fermion field, we here use the following fermion field, that is so-called the ringed fermion field [5],

$$\mathring{\chi}_f(t,x) = \sqrt{\frac{-2DN_f}{(4\pi)^2t^2\langle\bar{\chi}_f(t,x)\gamma_\mu \overleftrightarrow{D}_\mu\chi_f(t,x)\rangle_0}}\chi_f(t,x) = \chi_{f\,R}(t,x) + O(g^2). \tag{4.160}$$

The above discussion is a natural generalization to full QCD from the pure YM gauge theory. In fact, numerical implementations on the lattice have been already investigated via the studies on the thermodynamic quantities although only $t \to 0$ limit is taken with fine lattice spacing [31, 32]. The two-loop coefficients $c_i(t)$ are also discussed in Ref. [29].

References

1. Narayanan R, Neuberger H (2006) Infinite N phase transitions in continuum Wilson loop operators. JHEP 0603:064. arXiv:hep-th/0601210
2. Lüscher M (2010) Properties and uses of the Wilson flow in lattice QCD. JHEP 1008:071. arXiv:1006.4518 [hep-lat]
3. Lüscher M, Weisz P (2011) Perturbative analysis of the gradient flow in non-abelian gauge theories. JHEP 1102:051. arXiv:1101.0963 [hep-th]
4. Suzuki H (2013) Energy-momentum tensor from the Yang-Mills gradient flow. PTEP 2013(8):083B03. [Erratum: (2015) PTEP 2015(7):079201]. arXiv:1304.0533 [hep-lat]
5. Makino H, Suzuki H (2014) Lattice energy–momentum tensor from the Yang–Mills gradient flow—inclusion of fermion fields. PTEP 2014:063B02. [Erratum: (2015) PTEP 2015:079202]. arXiv:1403.4772 [hep-lat]
6. Albanese M et al (1987) [APE Collaboration] Glueball masses and string tension in lattice QCD. Phys Lett B 192:163
7. Morningstar C, Peardon MJ (2004) Analytic smearing of SU(3) link variables in lattice QCD. Phys Rev D 69:054501. arXiv:hep-lat/0311018 [hep-lat]
8. Hasenfratz A, Hoffmann R, Knechtli F (2002) The Static potential with hypercubic blocking. Nucl Phys B Proc Suppl 106:418–420. arXiv:hep-lat/0110168 [hep-lat]
9. Schaefer S et al (2011) [ALPHA Collaboration] Critical slowing down and error analysis in lattice QCD simulations. Nucl Phys B 845:93. arXiv:1009.5228 [hep-lat]
10. Lüscher M (2010) Trivializing maps, the Wilson flow and the HMC algorithm. Commun Math Phys 293:899–919. arXiv:0907.5491 [hep-lat]
11. Engel GP, Schaefer S (2011) Testing trivializing maps in the Hybrid Monte Carlo algorithm. Comput Phys Commun 182:2107–2114. arXiv:1102.1852 [hep-lat]
12. Asakawa M, Hatsuda T, Iritani T, Itou E, Kitazawa M, Suzuki H (2015) Determination of reference scales for wilson gauge action from Yang–Mills gradient flow. arXiv:1503.06516 [hep-lat]
13. Matsumoto T, Kitazawa M, Kohno Y (2021) Classifying topological charge in SU(3) Yang-Mills theory with machine learning. PTEP 2021(2):023D01. arXiv:1909.06238 [hep-lat]
14. Asakawa M, et al. (2014) [FlowQCD] Thermodynamics of SU(3) gauge theory from gradient flow on the lattice. Phys Rev D 90(1):011501. [Erratum: (2015) Phys Rev D 92(5):059902]. arXiv:1312.7492 [hep-lat]
15. Kitazawa M, Iritani T, Asakawa M, Hatsuda T, Suzuki H (2016) Equation of state for SU(3) gauge theory via the energy-momentum tensor under gradient flow. Phys Rev D 94(11):114512. arXiv:1610.07810 [hep-lat]
16. Iritani T, Kitazawa M, Suzuki H, Takaura H (2019) Thermodynamics in quenched QCD: energy–momentum tensor with two-loop order coefficients in the gradient-flow formalism. PTEP 2019(2):023B02. arXiv:1812.06444 [hep-lat]
17. Boyd G, Engels J, Karsch F, Laermann E, Legeland C, Lutgemeier M, Petersson B (1996) Thermodynamics of SU(3) lattice gauge theory. Nucl Phys B 469:419–444. arXiv:hep-lat/9602007 [hep-lat]
18. Borsanyi S, Endrodi G, Fodor Z, Katz SD, Szabo KK (2012) Precision SU(3) lattice thermodynamics for a large temperature range. JHEP 07:056. arXiv:1204.6184 [hep-lat]

19. Giusti L, Pepe M (2017) Equation of state of the SU(3) Yang-Mills theory: A precise determination from a moving frame. Phys Lett B 769:385–390. arXiv:1612.00265 [hep-lat]
20. Caselle M, Nada A, Panero M (2018) QCD thermodynamics from lattice calculations with nonequilibrium methods: the SU(3) equation of state. Phys Rev D 98(5):054513. arXiv:1801.03110 [hep-lat]
21. Hieda K, Makino H, Suzuki H (2017) Proof of the renormalizability of the gradient flow. Nucl Phys B 918:23–51. arXiv:1604.06200 [hep-lat]
22. Joglekar SD, Lee BW (1976) General theory of renormalization of gauge invariant operators. Ann Phys 97:160
23. Giusti L, Meyer HB (2011) Thermal momentum distribution from path integrals with shifted boundary conditions. Phys Rev Lett 106:131601. arXiv:1011.2727 [hep-lat]
24. Giusti L, Meyer HB (2013) Implications of Poincare symmetry for thermal field theories in finite-volume. JHEP 01:140. arXiv:1211.6669 [hep-lat]
25. Giusti L, Pepe M (2015) Energy-momentum tensor on the lattice: nonperturbative renormalization in Yang-Mills theory. Phys Rev D 91:114504. arXiv:1503.07042 [hep-lat]
26. Dalla Brida M, Giusti L, Pepe M (2020) Non-perturbative definition of the QCD energy-momentum tensor on the lattice. JHEP 04:043. arXiv:2002.06897 [hep-lat]
27. Caracciolo S, Curci G, Menotti P, Pelissetto A (1988) The energy momentum tensor on the lattice: the scalar case. Nucl Phys B 309:612
28. Caracciolo S, Curci G, Menotti P, Pelissetto A (1990) The Energy Momentum Tensor for Lattice Gauge Theories. Annals Phys 197:119
29. Harlander RV, Kluth Y, Lange F (2018) The two-loop energy–momentum tensor within the gradient-flow formalism. Eur Phys J C 78(11):944. [Erratum: (2019) Eur Phys J C 79(10):858]. arXiv:1808.09837 [hep-lat]
30. Lüscher M (2013) Chiral symmetry and the Yang-Mills gradient flow. JHEP 04:123. arXiv:1302.5246 [hep-lat]
31. Taniguchi Y, Ejiri S, Iwami R, Kanaya K, Kitazawa M, Suzuki H, Umeda T, Wakabayashi N (2017) Exploring $N_f = 2+1$ QCD thermodynamics from the gradient flow. Phys Rev D 96(1):014509. [Erratum: (2019) Phys Rev D 99(5):059904]. arXiv:1609.01417 [hep-lat]
32. Taniguchi Y, et al (2020) [WHOT-QCD] $N_f = 2+1$ QCD thermodynamics with gradient flow using two-loop matching coefficients. Phys Rev D 102(1):014510. arXiv:2005.00251 [hep-lat]

Chapter 5
Distribution of Energy Momentum Tensor around Single Static Quark in Deconfined Phase of SU(3) Yang–Mills Theory

To study complex quantum systems such as the Yang–Mills (YM) theory, it is customary to introduce test probes and analyze the response. In the previous studies based on numerical simulations on the lattice, in order to explore static quark and anti-quark systems, the properties of the potential (free energy) at zero (nonzero) temperature is analyzed, and the distributions of the action density and the magnitude of the color electric field are employed as physical quantities [1–16]. In these studies, however, our insights are restricted to global properties of the systems and/or somewhat indefinite physical meanings.

Compared with these studies, we employ a more direct observable, the energy momentum tensor (EMT) and investigate its distribution around static source(s). In fact, EMT is gauge invariant and an observable having a definite physical meaning related to energy density and force acting on a surface. In particular, principal directions of the stress tensor can serve as a gauge invariant definition of the direction of 'line of force' in non-Abelian theories. Moreover, EMT is a renormalization group invariant quantity, and its absolute value has an unambiguous meaning. In addition, as EMT is a second-rank tensor having many channels compared with a scalar/vector field, it provides us more detailed information on the system than the color electric field.

Thanks to the recent development of the gradient-flow method [17–19] and its application to EMT, $\mathcal{T}_{\mu\nu}(x)$ [20–23], it became possible to measure EMT even on the lattice, where the definition of EMT has been difficult due to the lack of the explicit translational invariance. The studies with EMT defined via the gradient flow on the lattice has succeeded in studying thermodynamics of YM theory [24–28] and of QCD [29, 30], which show the validity and the usefulness of the method through numerical simulations with reasonable statistics. These studies with EMT via the gradient flow focused on bulk quantities such as the pressure, energy density, entropy density, and the specific heat.

R. Yanagihara, *Distribution of Energy Momentum Tensor around Static Charges in Lattice Simulations and an Effective Model*, Springer Theses,
https://doi.org/10.1007/978-981-16-6234-8_5

Now we fully enjoy the viewpoint of the local interaction which EMT gives us. The purpose of this chapter is to explore the EMT distribution around a static quark (Q) in YM theory. As a first step, we consider the deconfined phase above the critical temperature T_c of the SU(3) YM theory in the range of temperature $1.2 \leq T/T_c \leq 2.6$ and measure the EMT distribution around the Polyakov loop. We focus on the local observable in this study and examine the following questions: (i) How are the energy density and the stress tensor distributed around the static quark? (ii) how are the distributions modified as a function of temperature? and (iii) how can one extract parameters such as the running coupling and the Debye screening mass from the distributions?

The present chapter is organized in the following way. First, in Sect. 5.1, we introduce the lattice observable, making a remark on the double extrapolation, $a \to 0$ followed by $t \to 0$. Then, we proceed to the numerical procedure with the lattice setup in Sect. 5.2. Numerical results are given in the final Sect. 5.3.

5.1 EMT around a Static Quark in SU(3) Yang–Mills Theory on the Lattice

5.1.1 Correlation

To describe a static Q on the lattice, we introduce the Polyakov loop at the origin, $\Omega(\mathbf{0})$. Then the expectation value of the EMT operator around Q is given by

$$\langle \mathcal{T}_{\mu\nu}(t, x)\rangle_Q = \frac{\langle \mathcal{T}_{\mu\nu}(t, x)\mathrm{Tr}\Omega(\mathbf{0})\rangle}{\langle \mathrm{Tr}\Omega(\mathbf{0})\rangle} - \langle \mathcal{T}_{\mu\nu}(t, x)\rangle. \tag{5.1}$$

Here we have two remarks with respect to Eq. (5.1). First, the EMT operator is now defined by

$$\mathcal{T}_{\mu\nu}(t, x) = c_1(t)U_{\mu\nu}(t, x)x + 4c_2(t)\delta_{\mu\nu}\left[E(t, x) - \langle E(t, x)\rangle_0\right], \tag{5.2}$$

instead of Eq. (4.143). The coefficients $c_1(t)$ and $c_2(t)$ represent two-loop order ones in perturbative calculation [23, 27] although the notation of the coefficients $\alpha_U(t)$ and $\alpha_E(t)$ is assigned for one-loop-order accuracy [20].

Second, we note that Eq. (5.1) is well defined only when the Z_3 symmetry in the SU(3) YM theory is spontaneously broken: In the Z_3 unbroken phase, both the numerator and denominator of the first term on the right-hand side vanish exactly. This is the reason why we focus on the system in the Z_3 broken phase above T_c in this study. In practice, we choose the state with the Polyakov loop being real among the three equivalent Z_3 states in the deconfined phase.

The renormalized EMT distribution around Q is obtained after taking the double extrapolation,

$$\langle \mathcal{T}^{\mathrm{R}}_{\mu\nu}(x)\rangle_Q = \lim_{t\to 0}\lim_{a\to 0}\langle \mathcal{T}_{\mu\nu}(t,x)\rangle_Q. \tag{5.3}$$

In our actual analysis, we extract the renormalized EMT distribution by fitting the lattice data with the following functional form [25, 26]:

$$\langle \mathcal{T}_{\mu\nu}(t,x)\rangle_Q = \langle \mathcal{T}^{\mathrm{R}}_{\mu\nu}(x)\rangle_Q + b_{\mu\nu}(t)a^2 + c_{\mu\nu}t + d_{\mu\nu}t^2, \tag{5.4}$$

where the contributions from discretization effects ($b_{\mu\nu}$) as well as the dimension-six and -eight operators ($c_{\mu\nu}$ and $d_{\mu\nu}$) are considered.

To perform the double extrapolation reliably, the smearing radius $\rho \equiv \sqrt{2t}$ needs to be larger than the lattice spacing to suppress the discretization error. At the same time, ρ should be smaller than half the temporal size $1/2T$ with temperature T as well as the distance from the source, r to avoid the overlap of operators. Therefore, we require

$$a/2 \lesssim \rho \lesssim \min\left(r, \frac{1}{2T}\right). \tag{5.5}$$

The lattice data to be fitted by Eq. (5.4) should be within this window. As will be discussed in Sect. 5.2.3, we impose more stringent conditions for the range of t in our numerical analysis.

5.1.2 Spherical Coordinate

For a system with a single static source, it is convenient to use the spherical coordinate system (r, θ, φ) with the radial coordinate $r = |\boldsymbol{x}|$, and the polar and azimuthal angles θ and φ. The spherical symmetry allows us to diagonalize the static EMT in Euclidean spacetime $\mathcal{T}_{\mu\nu}(\boldsymbol{x})$ in this coordinate system as

$$\mathcal{T}_{\gamma\gamma'}(\boldsymbol{x}) = \mathrm{diag}(\mathcal{T}_{44}(r), \mathcal{T}_{rr}(r), \mathcal{T}_{\theta\theta}(r)) \tag{5.6}$$

where $\gamma, \gamma' = 4, r, \theta$. Due to the spherical symmetry, the azimuthal component degenerates with the polar component, $\mathcal{T}_{\varphi\varphi}(r) = \mathcal{T}_{\theta\theta}(r)$, so that only independent components are given in Eq. (5.6).

In the Abelian case, EMT is given by the Maxwell stress-energy tensor [31], $\mathcal{T}^{\mathrm{Maxwell}}_{\mu\nu} = F_{\mu\rho}F_{\nu\rho} - \frac{1}{4}\delta_{\mu\nu}F_{\rho\sigma}F_{\rho\sigma}$ with the field strength $F_{\mu\nu}$. When a static electric charge is placed at the origin, the EMT is denoted by

$$\mathcal{T}^{\mathrm{Maxwell}}_{\gamma\gamma'} = \frac{1}{2}\mathrm{diag}(-\mathbf{E}^2, -\mathbf{E}^2, \mathbf{E}^2), \tag{5.7}$$

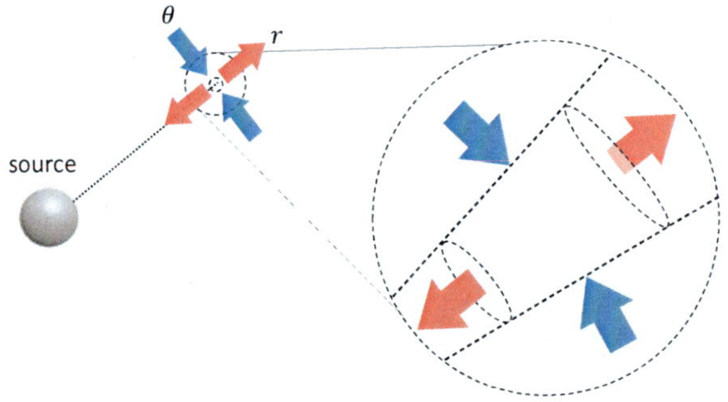

Fig. 5.1 Stress acting on infinitely small volume element under the existence of a single static charge (source). In the case of the classical electromagnetism, a small volume element at a distance of r from the charge is pulled along the radial direction from the neighboring volume elements while it is pushed in the transverse direction

with $E_i(\boldsymbol{x})$ being the electric field. The spatial structure of Eq. (5.7) is illustrated in Fig. 5.1 where the neighboring volume elements around the static electric charge pull (push) each other along the radial (angular) direction. In a static system, the force acting on a volume element through its surface should be balanced. This property is guaranteed by the momentum conservation $\partial_i \mathcal{T}_{ij} = 0$ together with the Gauss theorem.

5.2 Lattice Setup

5.2.1 *Gauge Configurations*

Numerical simulations in the SU(3) YM theory were performed on the four-dimensional Euclidean lattice with the Wilson gauge action and the periodic boundary conditions at four different temperatures $1.20T_c$, $1.44T_c$, $2.00T_c$, and $2.60T_c$. The simulation parameters for each T are summarized in Table 5.1. The inverse coupling $\beta = 6/g_0^2$ is related to the lattice spacing a determined by the reference scale w_0 [25, 32].[1] The spatial and temporal lattice sizes, N_s and N_τ together with the number of configurations N_{conf}, are also summarized in Table 5.1. All lattices have the same aspect ratio $N_s/N_\tau = 4$. The numerical simulations were carried out on OCTO-

[1] Throughout our studies, we employ the w_0-scale, which is the relation between the bare coupling or β and the lattice spacing a [25, 32]:

$$w_0 = 0.1670\,(\mathrm{fm}),$$
$$\frac{w_0}{a} = \exp\left(\frac{4\pi^2}{33}\beta - 9.1268 + \frac{41.806}{\beta} - \frac{158.26}{\beta^2}\right) \quad (6.3 \le \beta \le 7.4).$$

The conversion into the physical unit is based on this formula.

Table 5.1 Simulation parameters for the four temperatures: The spatial lattice size N_s, the temporal lattice size N_τ, $\beta = 6/g_0^2$, the lattice spacing a. $N_{\rm conf}$ represents the number of configurations

T/T_c	N_s	N_τ	β	a (fm)	$N_{\rm conf}$
1.20	40	10	6.336	0.0551	500
	48	12	6.467	0.0460	650
	56	14	6.581	0.0394	840
	64	16	6.682	0.0344	1,000
	72	18	6.771	0.0306	1,000
1.44	40	10	6.465	0.0461	500
	48	12	6.600	0.0384	650
	56	14	6.716	0.0329	840
	64	16	6.819	0.0288	1,000
	72	18	6.910	0.0256	1,000
2.00	40	10	6.712	0.0331	500
	48	12	6.853	0.0275	650
	56	14	6.973	0.0236	840
	64	16	7.079	0.0207	1,000
	72	18	7.173	0.0184	1,000
2.60	40	10	6.914	0.0255	500
	48	12	7.058	0.0212	650
	56	14	7.182	0.0182	840
	64	16	7.290	0.0159	1,000
	72	18	7.387	0.0141	1,000

PUS at the Cybermedia Center, Osaka University and Reedbush-U at Information Technology Center, The University of Tokyo.

The gauge configurations are generated by the pseudo-heat-bath method followed by five over-relaxations. Each measurement is separated by 200 sweeps, where we refer to the mixture of one heat bath update with five over-relaxation as 'sweep.' Statistical errors are estimated by the jackknife method with 20 jackknife bins. We employ the Wilson gauge action $S_{\rm W}(t)$ for $S(t)$ in the flow equation (4.48) and the clover-type representation for the field strength $G_{\mu\nu}(t,x)$. The numerical solution of the flow equation is obtained by the third-order Runge–Kutta method.

In order to suppress the statistical noise, we apply the multihit procedure in the measurement of the Polyakov loop by replacing every temporal link by its thermal average [33]. Details of the multihit algorism are summarized in Appendix A. The choice of the temporal argument x_4 of EMT in Eq. (5.1) is arbitrary. Therefore, we average EMT over the temporal direction to reduce the statistical error.

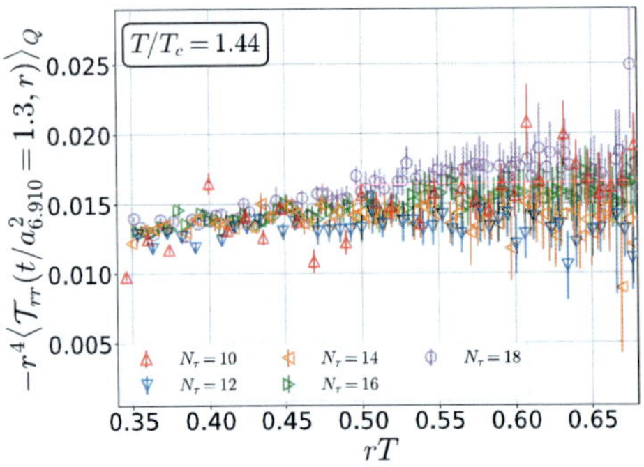

Fig. 5.2 Distribution of $-r^4\langle \mathcal{T}_{rr}(t/a_{6.910}^2 = 1.3, r)\rangle_Q$ as functions of rT at $T/T_c = 1.44$

5.2.2 Discretization Effect

The EMT in the spherical coordinate system on the lattice reads

$$T\sum_{x_4}\langle \mathcal{T}_{\gamma\gamma'}(t, \boldsymbol{x}, x_4)\rangle_Q = \mathrm{diag}(\langle \mathcal{T}_{44}(t,r)\rangle_Q, \langle \mathcal{T}_{rr}(t,r)\rangle_Q, \langle \mathcal{T}_{\theta\theta}(t,r)\rangle_Q). \quad (5.8)$$

The behavior of the EMT distribution close to the source Q is affected by the violation of rotational symmetry owing to lattice discretization. As an example, we show in Fig. 5.2 the distribution of $-r^4\langle \mathcal{T}_{rr}(t/a_{6.910}^2 = 1.3, r)\rangle_Q$ as a function of rT at $T/T_c = 1.44$, where $a_{6.910}$ is the lattice spacing of the finest lattice at this temperature. The figure shows that the oscillating behavior of the numerical results becomes more prominent on coarser lattices. In this study, we use the lattice data only for $N_\tau \geq 12$ for the continuum extrapolation to suppress the discretization errors. In Appendix B, we consider an alternative analysis that performs the tree-level improvement of the numerical results and uses them for the continuum extrapolation with the $N_\tau = 10$ data. As discussed there, we confirm that the results in both cases are the same within the errors.

5.2.3 Double Extrapolation

Although, in Chap. 4, we show the essential idea of the double extrapolation, $a \to 0$ followed by $t \to 0$, to obtain physical quantities, we consider the specific procedure such as the determination of the fitting range by using the lattice data at $T/T_c = 1.44$ as an example.

In Fig. 5.3, we show $-\langle \mathcal{T}_{rr}(t, rT)\rangle_Q/T^4$ at $rT = 0.40, 0.48, 0.60$ as a function of $1/N_\tau^2 = (aT)^2$ for four values of tT^2. To obtain the EMT values at a given r and t, we first perform the linear interpolation of the lattice data along the r direction and then interpolate along the t direction by the cubic spline method for each a.

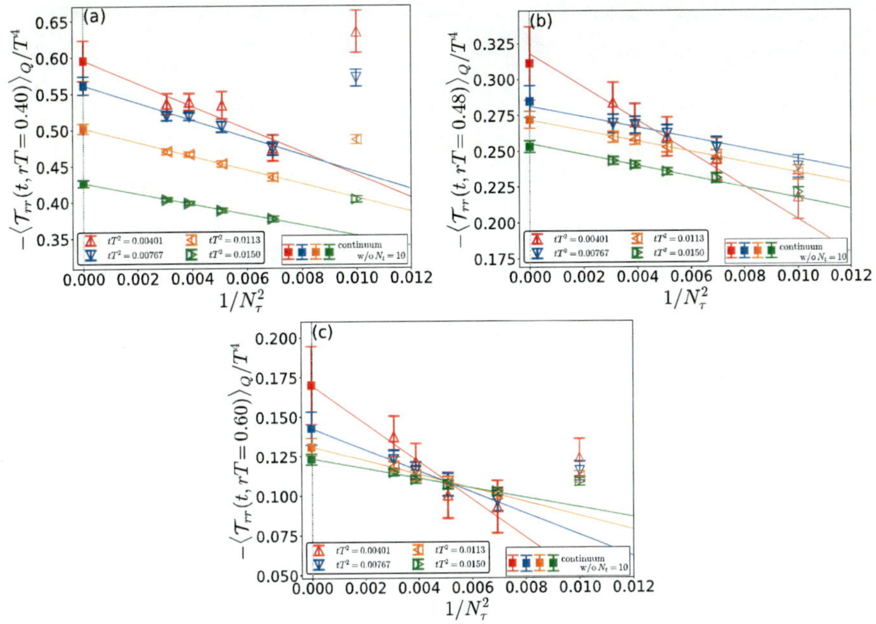

Fig. 5.3 Color open symbols represent $-\langle \mathcal{T}_{rr}(t, rT)\rangle_Q/T^4$ for various t as functions of $1/N_\tau^2 = a^2T^2$ at $T/T_c = 1.44$. The continuum extrapolation for each t is shown by the solid lines, with the extrapolated results represented by the filled symbols. Panels (**a**), (**b**) and (**c**) show the results at $rT = 0.40, 0.48, 0.60$, respectively

In Fig. 5.3, fitting results of the data at $N_\tau = 12$–18 according to Eq. (5.4) at fixed t are shown by the solid lines, while the results of the continuum limit are shown by the filled squares on the vertical dotted line at $1/N_\tau^2 = 0$. We then take $t \to 0$ extrapolation by fitting the continuum-extrapolated results for different t with Eq. (5.4) at $a = 0$. This fit has to be carried out within the range of t satisfying Eq. (5.5). We employ $tT^2 = 0.00401$ corresponding to $t/a^2 = 1.3$ of the finest lattice data as the lower bound of the fitting window: This choice satisfies Eq. (5.5) for all the lattices. The upper bound of the fitting window is taken to be $tT^2 = 0.015$, since the thermodynamic quantities show a linear behavior below this value [25].

We consider the following three ranges within $0.00401 \le tT^2 \le 0.015$ to estimate the systematic uncertainty from the fitting ranges [25]:

$$\text{Range-1} : 0.00767 \le tT^2 \le 0.0113,$$
$$\text{Range-2} : 0.00401 \le tT^2 \le 0.0113,$$
$$\text{Range-3} : 0.00767 \le tT^2 \le 0.0150.$$

Range-1 is the most conservative window, while Range-2 (Range-3) is the extension of Range-1 toward the smaller (larger) values of t. We employ the result of Range-1 as a central value and use the Range-2 and Range-3 for an estimate of the systematic

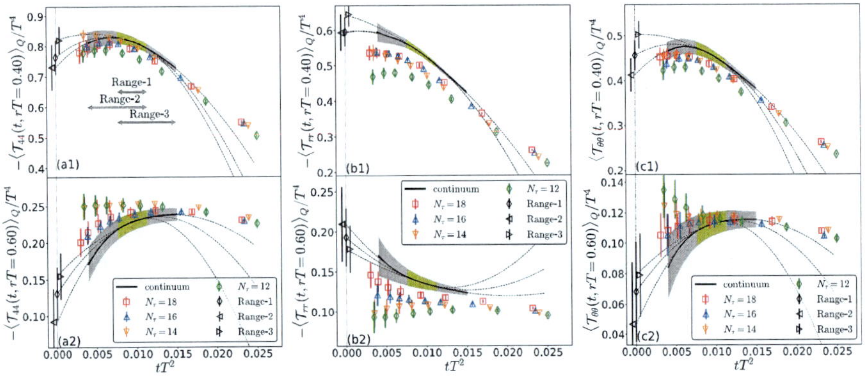

Fig. 5.4 Each component of EMT at $rT = 0.40$ (top) and $rT = 0.60$ (bottom) as functions of tT^2. Color open symbols denote $\langle \mathcal{T}_{\gamma\gamma'}(t, rT)\rangle_Q/T^4$ for each a as functions of tT^2. The black solid line with the gray error band is the continuum-extrapolated result. The dotted lines show the fitted results of the continuum result with Ranges-1, 2, and 3. The black symbols are the results of the $t \to 0$ extrapolation for these fitting ranges

error. In the following, all the results after the double extrapolation contain both the statistical and systematic errors.

In Fig. 5.4, the open symbols with statistical errors represent $\langle \mathcal{T}_{\gamma\gamma'}(t, rT)\rangle_Q/T^4$ at $rT = 0.40$ (upper) and 0.60 (lower) for each a. The results of the continuum limit are denoted by the black solid lines with the gray statistical error band for $0.00401 \le tT^2 \le 0.015$. Range-1 is highlighted by the yellow band. The figure also shows the fitted results for Ranges-1, 2, and 3 by the dotted lines. The final results of the $t \to 0$ limit for each range are shown by the open black symbols around $tT^2 = 0$: They agree with each other within the statistical errors, which suggests that the systematic uncertainty from the choice of the fitting range is not significant.

5.3 Results of EMT Distributions

Before entering into the detailed discussions on the spatial distribution of EMT, we first show the result of the stress distribution at $T/T_c = 2.60$ on a two-dimensional plane including the static source in Fig. 5.5. The same result is later shown in Fig. 5.8 in a different form. In Fig. 5.5, the red and blue arrows represent the principal directions of the stress tensor along the radial and transverse directions, respectively. The length of each arrow the represents square root of the eigenvalue corresponding to each principal axis. This figure is to be compared with Fig. 6.7 in Chap. 6.

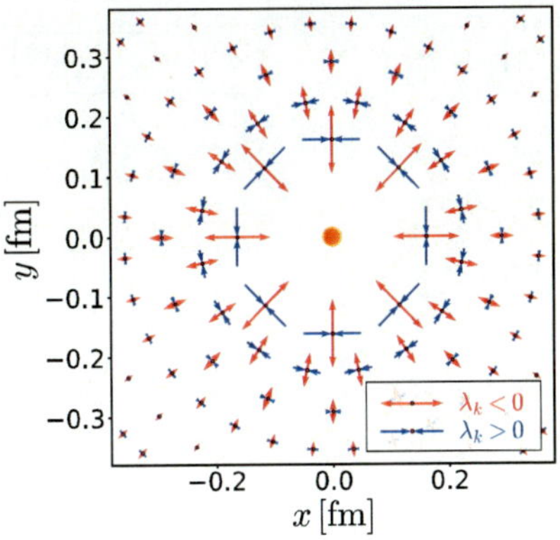

Fig. 5.5 Stress distribution around a static quark at the origin at $T/T_c = 2.60$. The red and blue arrows are the principal directions along the radial and transverse directions, respectively. The length of each arrow the represents square root of the eigenvalue

5.3.1 Channel Dependence

In Fig. 5.6, we show the dimensionless EMT, $-\langle \mathcal{T}^{\mathrm{R}}_{44}(r)\rangle_Q/T^4$, $-\langle \mathcal{T}^{\mathrm{R}}_{rr}(r)\rangle_Q/T^4$, and $\langle \mathcal{T}^{\mathrm{R}}_{\theta\theta}(r)\rangle_Q/T^4$, as functions of the dimensionless length rT. The error bands include both the statistical and systematic errors, where the latter is estimated from the three fitting ranges for the $t \to 0$ extrapolation. Since the thermal expectation value $\langle \mathcal{T}_{\mu\nu}(t, x)\rangle$ is subtracted as in Eq. (5.1), we have $\langle \mathcal{T}^{\mathrm{R}}_{\mu\nu}(r)\rangle_Q \to 0$ in the $r \to \infty$ limit.

We find that $-\langle \mathcal{T}^{\mathrm{R}}_{44}(r)\rangle_Q$, $-\langle \mathcal{T}^{\mathrm{R}}_{rr}(r)\rangle_Q$, and $\langle \mathcal{T}^{\mathrm{R}}_{\theta\theta}(r)\rangle_Q$ are all positive for $rT \lesssim 1$ and decrease rapidly with increasing r. These signs are the same as those of the Maxwell stress tensor in Eq. (5.7). Individual signs physically mean that a volume element has a positive localized energy density and receives a pulling (pushing) force along the longitudinal (transverse) direction.

Figure 5.6 indicates that the absolute values of the spatial components $|\langle \mathcal{T}^{\mathrm{R}}_{rr}(r)\rangle_Q|$ and $|\langle \mathcal{T}^{\mathrm{R}}_{\theta\theta}(r)\rangle_Q|$ are degenerated within the error for all temperatures. On the other hand, $|\langle \mathcal{T}^{\mathrm{R}}_{44}(r)\rangle_Q|$ is larger than the spatial components especially at lower temperature. This is in contrast to the degenerate magnitude of all components in the Maxwell stress Eq. (5.7).[2]

5.3.2 Temperature Dependence

Shown in Fig. 5.7 is the temperature dependence of the spatial distribution of EMT with respect to the physical distance r (fm); (a) $-\langle \mathcal{T}^{\mathrm{R}}_{44}(r)\rangle_Q$, (b) $-\langle \mathcal{T}^{\mathrm{R}}_{rr}(r)\rangle_Q$, and (c) $\langle \mathcal{T}^{\mathrm{R}}_{\theta\theta}(r)\rangle_Q$. Also, shown in Fig. 5.7(d) is the distribution of the trace of EMT given by

$$\Delta_Q(r) \equiv -\langle \mathcal{T}^{\mathrm{R}}_{\mu\mu}(r)\rangle_Q = -\langle \mathcal{T}^{\mathrm{R}}_{44}(r) + \mathcal{T}^{\mathrm{R}}_{rr}(r) + 2\mathcal{T}^{\mathrm{R}}_{\theta\theta}(r)\rangle_Q. \tag{5.9}$$

[2] See also Appendix C.

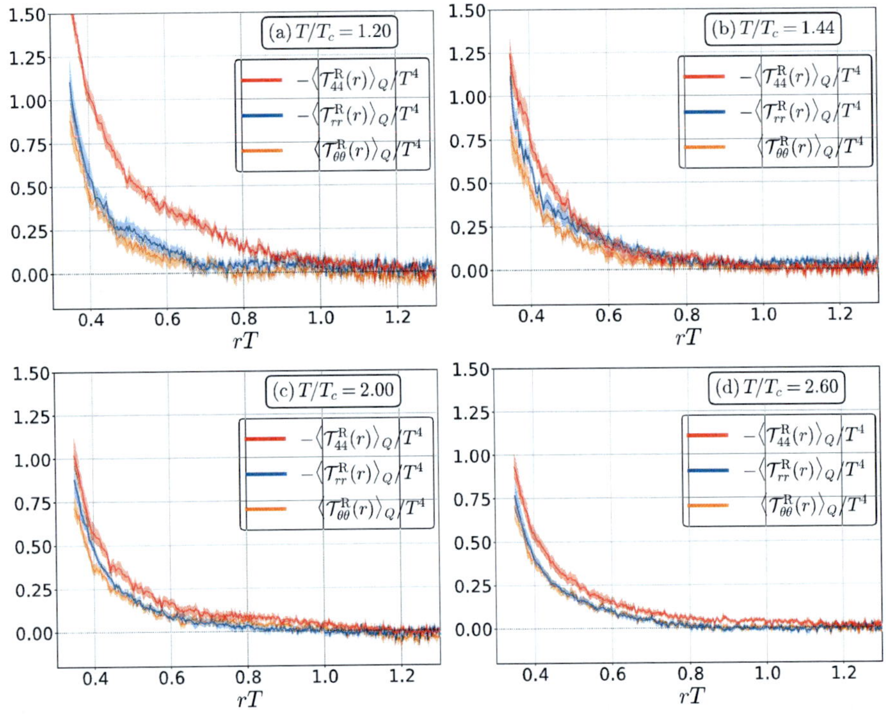

Fig. 5.6 EMT distribution $(-\langle \mathcal{T}^{\mathrm{R}}_{44}(r)\rangle_Q, -\langle \mathcal{T}^{\mathrm{R}}_{rr}(r)\rangle_Q, \langle \mathcal{T}^{\mathrm{R}}_{\theta\theta}(r)\rangle_Q)$ as functions of rT after the double extrapolation: (**a**) $T = 1.20T_c$, (**b**) $T = 1.44T_c$, (**c**) $T = 2.00T_c$, and (**d**) $T = 2.60T_c$

Figure 5.7 tells us that the EMT distributions have small T dependence at short distances, $r \lesssim 0.2$ fm. On the other hand, for large distances, sizable T dependence can be seen despite the growth of the errors at high T.

To make these features more explicit, we plot the same results with a dimensionless normalization $r^4\langle \mathcal{T}^{\mathrm{R}}_{\gamma\gamma'}(r)\rangle_Q$ as a function of r in Fig. 5.8. The figure shows that the T dependence is suppressed for $rT \lesssim 0.3$ and all results approach a single line, while the result tends to be more suppressed for $rT \gtrsim 0.3$ compared with this universal behavior as temperature is raised. This result is reasonable as the T dependence of $r^4\langle \mathcal{T}^{\mathrm{R}}_{\gamma\gamma'}(r)\rangle_Q$ would be suppressed for $r \lesssim (2\pi T)^{-1}$.

At distance $(2\pi T)^{-1} \ll r \ll \Lambda^{-1}$ with Λ being the lambda parameter, the behavior of $\langle \mathcal{T}^{\mathrm{R}}_{\gamma\gamma'}(r)\rangle_Q$ should be described by the perturbation theory in electrostatic QCD (EQCD). In the leading order of EQCD in this regime, we have the following ratio (Appendix C),

$$\left| \frac{\Delta_Q(r)}{\langle \mathcal{T}^{\mathrm{R}}_{44,rr,\theta\theta}(r)\rangle_Q} \right| = \frac{11}{2\pi}\alpha_s + O(g^3), \tag{5.10}$$

which is independent of r and T and is given only by a function of α_s.

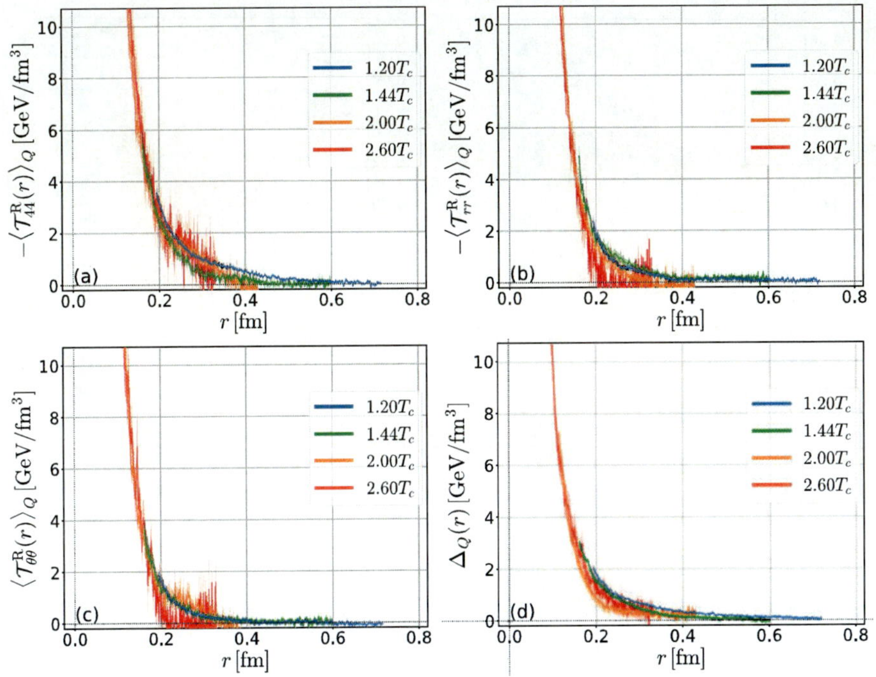

Fig. 5.7 EMT distribution $(-\langle \mathcal{T}^{\mathrm{R}}_{44}(r)\rangle_Q, -\langle \mathcal{T}^{\mathrm{R}}_{rr}(r)\rangle_Q, \langle \mathcal{T}^{\mathrm{R}}_{\theta\theta}(r)\rangle_Q)$ and $\Delta_Q(r)$ as functions of r (fm) at each temperature: (**a**) $-\langle \mathcal{T}^{\mathrm{R}}_{44}(r)\rangle_Q$, (**b**) $-\langle \mathcal{T}^{\mathrm{R}}_{rr}(r)\rangle_Q$, (**c**) $\langle \mathcal{T}^{\mathrm{R}}_{\theta\theta}(r)\rangle_Q$, and (**d**) $\Delta_Q(r)$

Shown in Fig. 5.9 is the r dependence of $\Delta_Q(r)/\langle \mathcal{T}^{\mathrm{R}}_{44,rr,\theta\theta}(r)\rangle_Q$ as a function of r at $T/T_c = 2.60$. From this result and Eq. (5.10), we obtain, at $r = 0.1$ fm, $\alpha_s = 0.221(17)$ from $-\Delta_Q(r)/\langle \mathcal{T}^{\mathrm{R}}_{44}(r)\rangle_Q$, $\alpha_s = 0.286(24)$ from $-\Delta_Q(r)/\langle \mathcal{T}^{\mathrm{R}}_{rr}(r)\rangle_Q$, and $\alpha_s = 0.319(35)$ from $\Delta_Q(r)/\langle \mathcal{T}^{\mathrm{R}}_{\theta\theta}(r)\rangle_Q$. Although these values are channel dependent, indicating the existence of non-negligible higher order contributions, it is notable that they are consistent with that obtained from the similar analysis of the Polyakov loop correlations at $r = 0.1$ fm [34, 35]. Higher-order α_s corrections and the thermal corrections for EMT around a static charge to be compared with our lattice data are expected.

Let us now turn to the long-distance region in Fig. 5.8. Owing to the large errors in this region, it is not possible to extract the thermal screening of the form $\exp(-2m_{\mathrm{D}}r)$ with m_{D} being the Debye screening mass. Nevertheless, Fig. 5.8(a)–(d) indicate that the EMT distributions decrease faster than $1/r^4$ at long distances, and the tendency is stronger at high temperatures. To draw a definite conclusion, however, higher statistics are necessary.

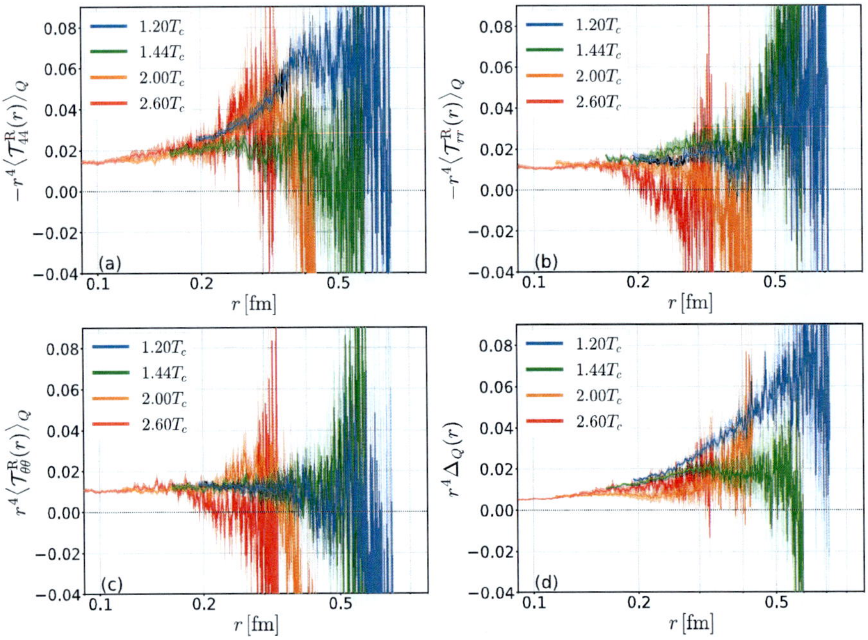

Fig. 5.8 EMT distribution $r^4(-\langle\mathcal{T}^{\rm R}_{44}(r)\rangle_Q, -\langle\mathcal{T}^{\rm R}_{rr}(r)\rangle_Q, \langle\mathcal{T}^{\rm R}_{\theta\theta}(r)\rangle_Q)$ and $r^4\Delta_Q(r)$ as functions of r (fm) at each temperature: **(a)** $-r^4\langle\mathcal{T}^{\rm R}_{44}(r)\rangle_Q$, **(b)** $-r^4\langle\mathcal{T}^{\rm R}_{rr}(r)\rangle_Q$, **(c)** $r^4\langle\mathcal{T}^{\rm R}_{\theta\theta}(r)\rangle_Q$, and **(d)** $r^4\Delta_Q(r)$

Fig. 5.9 Ratios $\Delta_Q(r)/\langle\mathcal{T}^{\rm R}_{44,rr,\theta\theta}(r)\rangle_Q$ as functions of r at $T/T_c = 2.60$

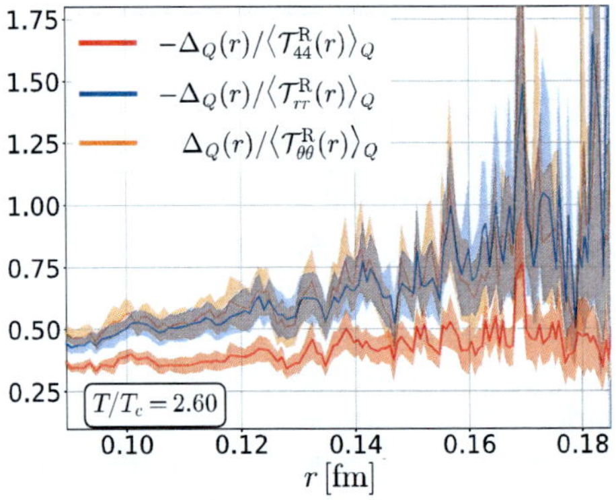

References

1. Bali GS (2001) QCD forces and heavy quark bound states. Phys Rep 343:1–136. arXiv:hep-ph/0001312 [hep-ph]
2. Bali GS, Schilling K (1993) Running coupling and the Lambda parameter from SU(3) lattice simulations. Phys Rev D 47:661–672. arXiv:hep-lat/9208028 [hep-lat]
3. Bali GS, Schilling K, Wachter A (1997) Complete $\mathcal{O}(v^2)$ corrections to the static interquark potential from SU(3) gauge theory. Phys Rev D 56:2566–2589. arXiv:hep-lat/9703019 [hep-lat]

4. Cardoso N, Cardoso M, Bicudo P (2013) Inside the SU(3) quark-antiquark QCD flux tube: screening versus quantum widening. Phys Rev D 88:054504. arXiv:1302.3633 [hep-lat]
5. Cea P, Cosmai L, Papa A (2012) Chromoelectric flux tubes and coherence length in QCD. Phys Rev D 86:054501. arXiv:1208.1362 [hep-lat]
6. Maezawa Y, Umeda T, Aoki S, Ejiri S, Hatsuda T, Kanaya K, Ohno H (2012) Application of fixed scale approach to static quark free energies in quenched and 2+1 flavor lattice QCD with improved Wilson quark action. Prog Theor Phys 128:955–970. arXiv:1112.2756 [hep-lat]
7. Di Giacomo A, Maggiore M, Olejnik S (1990) Confinement and Chromoelectric Flux Tubes in Lattice QCD. Nucl Phys B 347:441–460
8. Bali GS, Schilling K, Schlichter C (1995) Observing long color flux tubes in SU(2) lattice gauge theory. Phys Rev D 51:5165–5198. arXiv:hep-lat/9409005 [hep-lat]
9. Michael C (1996) Lattice sum rules for the color fields. Phys Rev D 53:4102–4105. arXiv:hep-lat/9504016 [hep-lat]
10. Green AM, Michael C, Spencer PS (1997) The Structure of flux tubes in SU(2). Phys Rev D 55:1216–1225. arXiv:hep-lat/9610011 [hep-lat]
11. Gliozzi F, Pepe M, Wiese UJ (2010) The width of the confining string in Yang-Mills theory. Phys Rev Lett 104:232001. arXiv:1002.4888 [hep-lat]
12. Meyer HB (2010) The transverse structure of the QCD string. Phys Rev D 82:106001. arXiv:1008.1178 [hep-lat]
13. Cea P, Cosmai L, Cuteri F, Papa A (2016) Flux tubes at finite temperature. JHEP 06:033 [arXiv:1511.01783 [hep-lat]]
14. Cea P, Cosmai L, Cuteri F, Papa A (2017) Flux tubes in the QCD vacuum. Phys Rev D 95(11):114511. arXiv:1702.06437 [hep-lat]
15. Baker M, Cea P, Chelnokov V, Cosmai L, Cuteri F, Papa A (2019) Isolating the confining color field in the SU(3) flux tube. Eur Phys J C 79(6):478. arXiv:1810.07133 [hep-lat]
16. Baker M, Cea P, Chelnokov V, Cosmai L, Cuteri F, Papa A (2020) The confining color field in SU(3) gauge theory. Eur Phys J C 80(6):514. arXiv:1912.04739 [hep-lat]
17. Narayanan R, Neuberger H (2006) Infinite N phase transitions in continuum Wilson loop operators. JHEP 0603:064. arXiv:hep-th/0601210
18. Lüscher M (2010) Properties and uses of the Wilson flow in lattice QCD. JHEP 1008:071. arXiv:1006.4518 [hep-lat]
19. Lüscher M, Weisz P (2011) Perturbative analysis of the gradient flow in non-abelian gauge theories. JHEP 1102:051. arXiv:1101.0963 [hep-th]
20. Suzuki H (2015) Energy-momentum tensor from the Yang-Mills gradient flow. PTEP 2013(8):083B03. [Erratum: (2015) PTEP 2015(7):079201]. arXiv:1304.0533 [hep-lat]
21. Makino H, Suzuki H (2014) Lattice energy–momentum tensor from the Yang–Mills gradient flow—inclusion of fermion fields. PTEP 2014:063B02. [Erratum: (2015) PTEP 2015:079202]. arXiv:1403.4772 [hep-lat]
22. Hieda K, Makino H, Suzuki H (2017) Proof of the renormalizability of the gradient flow. Nucl Phys B 918:23–51. arXiv:1604.06200 [hep-lat]
23. Harlander RV, Kluth Y, Lange F (2019) The two-loop energy–momentum tensor within the gradient-flow formalism. Eur Phys J C 78(11):944. [Erratum: (2019) Eur Phys J C 79(10):858]. arXiv:1808.09837 [hep-lat]
24. Asakawa M, et al (2014) [FlowQCD] Thermodynamics of SU(3) gauge theory from gradient flow on the lattice. Phys Rev D 90(1):011501. [Erratum: (2015) Phys Rev D 92(5):059902]. arXiv:1312.7492 [hep-lat]
25. Kitazawa M, Iritani T, Asakawa M, Hatsuda T, Suzuki H (2016) Equation of state for SU(3) Gauge theory via the energy-momentum tensor under gradient flow. Phys Rev D 94(11):114512. arXiv:1610.07810 [hep-lat]
26. Kitazawa M, Iritani T, Asakawa M, Hatsuda T (2017) Correlations of the energy-momentum tensor via gradient flow in SU(3) Yang-Mills theory at finite temperature. Phys Rev D 96(11):111502. arXiv:1708.01415 [hep-lat]
27. Iritani T, Kitazawa M, Suzuki H, Takaura H (2019) Thermodynamics in quenched QCD: energy–momentum tensor with two-loop order coefficients in the gradient-flow formalism. PTEP 2019(2):023B02. arXiv:1812.06444 [hep-lat]

28. Kitazawa M, Mogliacci S, Kolbé I, Horowitz WA (2019) Anisotropic pressure induced by finite-size effects in SU(3) Yang-Mills theory. Phys Rev D 99(9):094507. arXiv:1904.00241 [hep-lat]
29. Taniguchi Y, Ejiri S, Iwami R, Kanaya K, Kitazawa M, Suzuki H, Umeda T, Wakabayashi N (2017) Exploring N_f = 2+1 QCD thermodynamics from the gradient flow. Phys Rev D 96(1):014509. [Erratum: (2019) Phys Rev D 99(5):059904]. arXiv:1609.01417 [hep-lat]
30. Taniguchi Y, et al. (2020) [WHOT-QCD] N_f = 2+1 QCD thermodynamics with gradient flow using two-loop matching coefficients. Phys Rev D 102(1):014510. arXiv:2005.00251 [hep-lat]
31. Landau LD, Lifshitz EM (1980) The classical theory of fields, 4th edn. Butterworth-Heinemann
32. Borsanyi S, Durr S, Fodor Z, Hoelbling C, Katz SD, Krieg S, Kurth T, Lellouch L, Lippert T, McNeile C et al (2012) High-precision scale setting in lattice QCD. JHEP 09:010. arXiv:1203.4469 [hep-lat]
33. Parisi G, Petronzio R, Rapuano F (1983) A measurement of the string tension near the continuum limit. Phys Lett 128B:418
34. Kaczmarek O, Karsch F, Zantow F, Petreczky P (2004) Static quark anti-quark free energy and the running coupling at finite temperature. Phys Rev D 70:074505. [Erratum: (2005) Phys Rev D 72:059903]. arXiv:hep-lat/0406036 [hep-lat]
35. Kaczmarek O, Zantow F (2005) Static quark anti-quark interactions in zero and finite temperature QCD. I. Heavy quark free energies, running coupling and quarkonium binding. Phys Rev D 71:114510. arXiv:hep-lat/0503017 [hep-lat]

Chapter 6
Distribution of Energy Momentum Tensor around Static Quark–Anti-Quark in Vacuum of SU(3) Yang–Mills Theory

In the previous chapter, we explore the distribution of the energy momentum tensor (EMT) around a static quark at finite temperature above T_c of the SU(3) Yang–Mills (YM) theory on the lattice. From the resulting EMT distribution, we find that there exists a genuine non-Abelian effect at finite temperature.

On the basis of the same formalism as presented in the previous chapter, we now proceed to the exploration of the EMT distribution under the presence of static quark (Q) and anti-quark ($\bar{Q}$) charges separated at a distance R in the SU(3) YM theory in vacuum on the lattice. In such a setup, the YM field strength is believed to be squeezed into a quasi-one-dimensional flux tube [1–3] and gives rise to the linear confining potential at large R (see the reviews [4–6] and references therein), which indicates the quark confinement. This picture is in sharp contrast to the situation with two opposite charges in Maxwell theory, where both charges interact with each other through the Coulomb force and the electric field spreads all over the space.

Although the action density, the color electric field, and the plaquettes have been employed before to probe such a flux tube [7–14], the purpose of the present study is to attempt to provide gauge-invariant EMT distribution around the $Q\bar{Q}$ pair in three spatial dimensions. Since EMT can describe the local response to the distortion of the YM vacuum which is stemmed from external sources in a gauge-invariant manner, EMT prominently contributes to provide non-perturbative information especially on the quantum YM theory. The fundamental theoretical tool to make this analysis possible is the YM gradient flow [15–17].

This chapter is composed of the following sections. First, in Sect. 6.1, we consider the Wilson loop, which plays a role of creation of static $Q\bar{Q}$ and confirm the linear rising potential. Here, in order to improve a coupling of the ground state of $Q\bar{Q}$, we employ APE smearing method and determine an appropriate number of APE smearing. The details on the APE smearing method are summarized in Appendix D. In Sect. 6.2, we explain the quantity to measure and the useful coordinate, which is

R. Yanagihara, *Distribution of Energy Momentum Tensor around Static Charges in Lattice Simulations and an Effective Model*, Springer Theses,
https://doi.org/10.1007/978-981-16-6234-8_6

followed by an introduction of the setup of our numerical simulation in Sect. 6.3. Numerical results and their physical implications are given in the Sects. 6.4 and 6.5 after we present the procedure of the double extrapolation to obtain physical quantities.

6.1 Wilson Loop and $Q\bar{Q}$ Potential

In order to prepare $Q\bar{Q}$ on the lattice, we consider the standard Wilson loop in the SU(3) YM theory. The purpose of this section is to show the $Q\bar{Q}$ potential from the expectation value of the Wilson loop. In addition, we determine the appropriate number of APE smearing to effectively explore the $Q\bar{Q}$ system on the lattice. Although the procedure to relate the Wilson loop and the potential of the ground state follows, in principle, the discussion in Chap. 3, we perform the following improvements for practical calculations:

1. In order to enhance the coupling of the Wilson loop to the ground state in the $Q\bar{Q}$ system, we perform the spatial APE smearing [18], which plays a role to render the coefficient concerning the potential of the ground state dominant in the expectation value of the Wilson loop as mentioned in Appendix D. In our study, we employ the conventional smearing parameter $\alpha = 2.3$ [19].
2. As mentioned in Appendix A, we also perform the local update of link variables along the temporal direction by substituting their mean fields for the noise reduction in the measurement of the Wilson loop. This technique is referred to as the multihit procedure to reduce the statistical noise. The number of updates is 100 in our actual simulation, considering a balance between the simulation time and the saturation of the noise-to-signal (N/S) ratio [20].

Shown in the following, when the number of the iterations of the APE smearing becomes larger, the coupling of the Wilson loop to the $Q\bar{Q}$ ground state is enhanced. At the same time, however, the simulation error increases. Thus, it is suggested that the appropriate number of iteration in APE smearing exists. We determine it from the lattice simulation at $\beta = 6.600$, where β is related to the bare coupling as $\beta = 6/g_0^2$. All the simulation parameters are summarized in Table 6.1. N_{size}^4, N_{conf} and N_{APE} represent the lattice size, the number of statistics and the number of APE smearing, respectively. R/a and T/a are the spatial and temporal length of the Wilson loop.

The numerical simulations in the SU(3) YM gauge theory were performed on the four-dimensional Euclidean lattice with the Wilson gauge action and the periodic boundary condition. Gauge configurations were generated by the heat bath method with the over-relaxation. Each measurement is separated after 100 sweeps updates. The calculation was carried out on IBM System Blue Gene Solution at KEK. In the following, we focus on the analyses at $R/a = 12, 18, 24$ in particular.

Table 6.1 Simulation parameters for the determination of the appropriate number of iterations in APE smearing. $\beta = 6/g_0^2$ is related to the lattice spacing a. $N_{\rm size}^4$, $N_{\rm conf}$, and $N_{\rm APE}$ represent the lattice size, the number of statistics, and the number of APE smearing, respectively. R/a and T/a are the spatial and temporal length of the Wilson loop

β	a (fm)	$N_{\rm size}^4$	$N_{\rm conf}$	$N_{\rm APE}$	R/a	T/a
6.600	0.038	48^4	100	{80,160,240, 320,400}	$\{1, 2, \ldots, 33\}$	$\{1, 2, \ldots, 15\}$

6.1.1 $Q\bar{Q}$ Potential

As shown in Eq. (3.68), the Wilson loop has the contribution of not only the ground state but also states with the higher energies. In Eq. (3.68), the coefficients C_i ($i = 0, 1, \ldots$) depend on the $Q\bar{Q}$ distance R. Then the $Q\bar{Q}$ potential is, in principle, extracted by

$$V(R) \equiv V_0(R) = -\lim_{T\to\infty} \frac{1}{T} \ln\langle W_{C=R\times T}[U]\rangle. \tag{6.1}$$

In the actual analyses with the APE smearing, we consider the effective mass

$$V_{\rm eff}(R,T) = \ln \frac{\langle W(R,T)\rangle}{\langle W(R,T+a)\rangle} = V(R) + \frac{C_1(R)}{C_0(R)} h(R,T) + \cdots, \tag{6.2}$$

where

$$h(R,T) = e^{-(V_1(R)-V(R))T}(1 - e^{-(V_1(R)-V(R))}). \tag{6.3}$$

When the effective mass as a function of T with respect to each R has a plateau for large T, the ground state potential $V(R)$ can be extracted. Shown in Fig. 6.1 is the plot of the effective mass for various numbers of APE smearing at $R/a = 12, 18, 24$.

From Fig. 6.1, we confirm that there exist plateaux for $N_{\rm APE} = 160, 240$ at $T/a \geq 5$ although there does not for $N_{\rm APE} = 80$. The behavior of $V_{\rm eff}$ at small T is attributed to the insufficiency of the coupling to the ground state while that in the large T region suffers from the statistical fluctuation. We also find that large $N_{\rm APE}$ impedes the existence of the plateau. Therefore, it is indicated that there exists an appropriate number of APE smearing for the emergence of the plateau. Although the determination of $N_{\rm APE}$ is dependent on R, we consider that the condition $160 \leq N_{\rm APE} \leq 240$ at least ensures the plateau at $T/a \geq 5$ for all R. In our analysis to calculate the $Q\bar{Q}$ potential, we employ $N_{\rm APE} = 160$ and regard $V_{\rm eff}$ in $5 \leq T/a \leq 9$ as the plateau. In Fig. 6.1, the black horizontal line with gray band represents the mean value with the statistical error of the plateau. The statistical error is estimated by the jackknife method with the number of the jackknife bin is 20, which we have confirmed saturate the error.

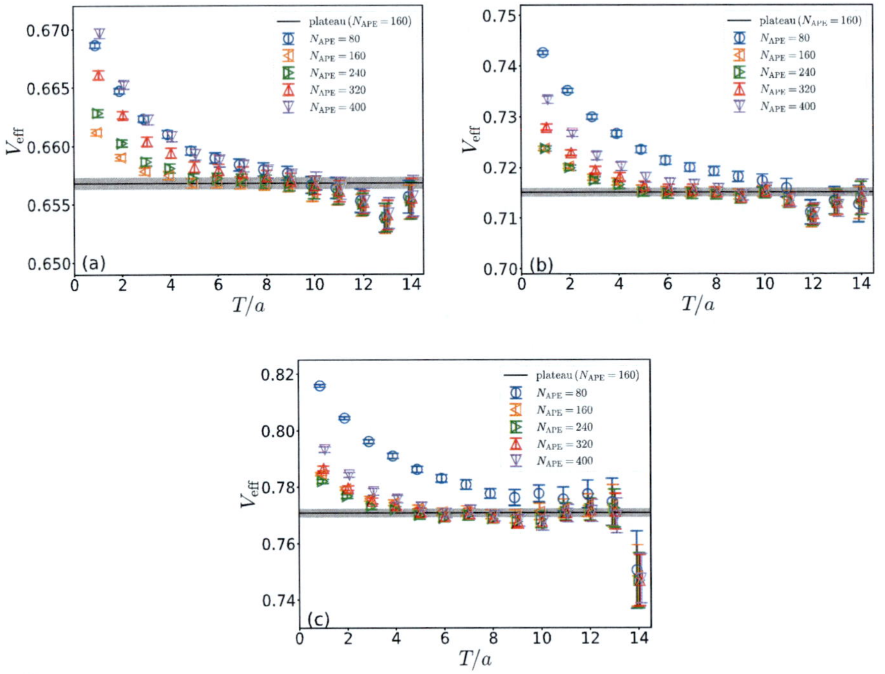

Fig. 6.1 Effective mass plot. From the left figure, (**a**) $R/a = 12$, (**b**) $R/a = 18$ and (**c**) $R/a = 24$

Figure 6.2 is the $Q\bar{Q}$ potential determined from the plateaux according to the above criteria. In the same figure, we fit the result with the Cornell parameterization

$$V_{\mathrm{Cornell}}(R) = -\frac{\mathcal{A}}{R} + \sigma R + \mathcal{B}, \tag{6.4}$$

which is typically used to describe the confinement potential. Here $\mathcal{A}$, σ, and $\mathcal{B}$ are the fitting parameters. We notice that we do not employ data at $R/a = 1, 2$ to fit the result due to the large discretization effect. In particular, it is strictly prohibited to use data at $R/a = 1$ because, in the multihit procedure for the links along the temporal direction in the Wilson loop, the staple is obliged to include the link in the opposite side.

Figure 6.2 shows the linearly confining behavior in the large R region, which suggests the confinement and σ characterizes the confining force. The noise reduction which we employ plays an adequate role to make errors small. We remark that there does not exist a finite volume effect since the result at $R > 24$, where $R/a = 24$ corresponds to half the volume $N_{\mathrm{size}} = 48$, still shows a linear growth.

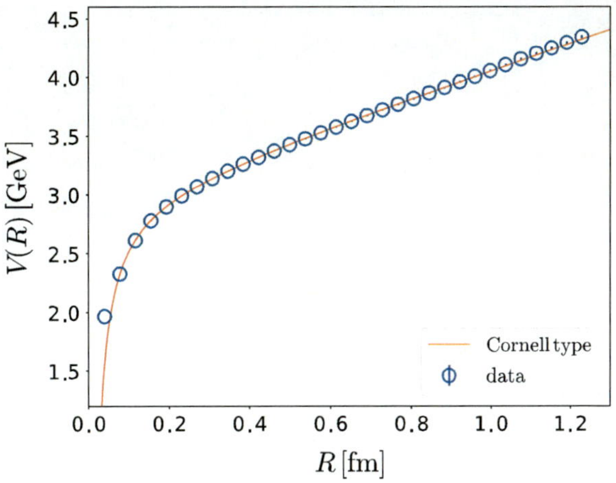

Fig. 6.2 $Q\bar{Q}$ potential

6.1.2 Determination of N_{APE}

From the consideration in the previous section, we find that there exists an appropriate range of N_{APE}. In this section, from the viewpoint of both how large the coupling to the ground state is and the N/S ratio of the Wilson loop is, we again confirm that $N_{\mathrm{APE}} = 160$ is one of the most appropriate choices.

The coupling of the Wilson loop to the ground state is defined by

$$R_{\mathrm{G}} = \frac{C_0(R)\exp[-V(R)T]}{\langle W(R,T)\rangle}, \tag{6.5}$$

and the N/S ratio of the Wilson loop is $\delta\langle W(R,T)\rangle / \langle W(R,T)\rangle$. Shown in Fig. 6.3 is the N/S ratio as a function of T/a at $R/a = 12, 18, 24$. Different colors correspond to each N_{APE}. All the data in Fig. 6.3 basically show the larger T and/or the larger N_{APE} increase the N/S ratio.

On the other hand, Fig. 6.4 corresponds to R_{G} as a function of T/a at $R/a = 12, 18, 24$. In this analysis, we obtain the coefficient $C_0(R)$ from the plateau of the following ratio:

$$C_0(R,T) = \frac{\langle W(R,T)\rangle^{T+a}}{\langle W(R,T+a)\rangle^{T}} = C_0(R) + C_1(R)h(R,T) + \cdots, \tag{6.6}$$

where $h(R,T)$ is defined by Eq. (6.3). Here we also employ $5 \le T/a \le 9$ as a plateau for the consistency with the above analysis. All the results for various N_{APE} represent increases of R_{G} with larger T/a although there are decreases at $T/a > 10$, which we consider to be attributed to the statistical fluctuations. We also find that R_{G} at fixed T becomes large for $N_{\mathrm{APE}} \lesssim 160$ when N_{APE} becomes large. For $N_{\mathrm{APE}} \gtrsim 160$, however, R_{G} becomes smaller with the increase of N_{APE}. This behavior results from the fact that N_{APE} corresponds to the physical transverse size of the flux tube, where N_{APE}

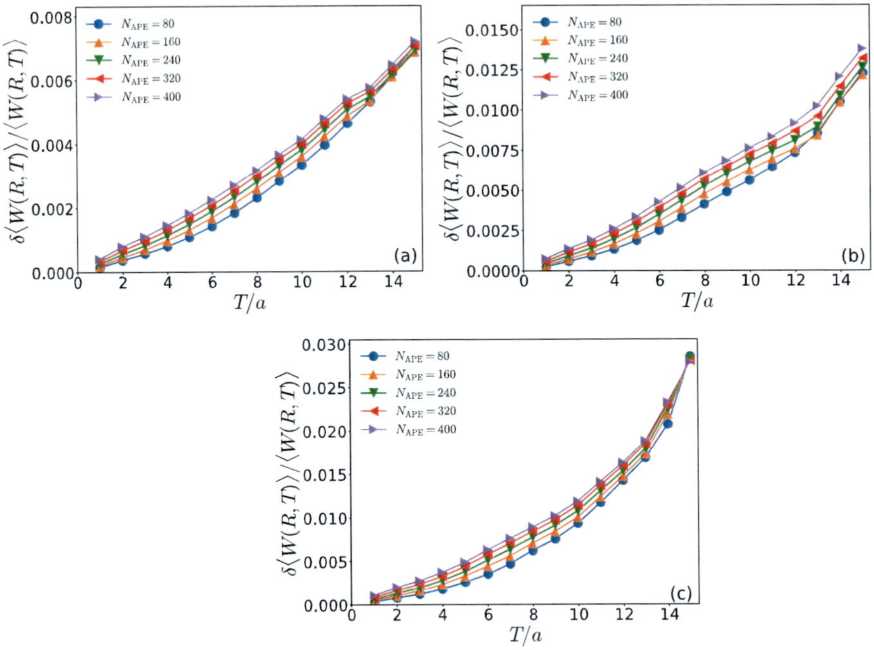

Fig. 6.3 N/S ratio of the Wilson loop as a function of T/a. From the left figure, (**a**) $R/a = 12$, (**b**) $R/a = 18$, and (**c**) $R/a = 24$

represents how 'fattened' links are [19]. Thus, the appropriate number of iterations in the APE smearing is suggested.

Figure 6.5 is a summary of $N_{\rm APE}$-dependence of $R_{\rm G}$ and the N/S ratio at $R/a = 12, 18, 24$. Here we fix the temporal size of the Wilson loop as $T/a = 5$, which is the minimum value of T in the plateau.

Therefore, based on these observations, we confirm that $N_{\rm APE} = 160$ is one of the most appropriate choices. Although we should vary $N_{\rm APE}$ depending on R, we employ the same value for all R at fixed β. We confirm that the coupling of the Wilson loop to the ground state sufficiently saturates such that $|1 - R_{\rm G}| < 0.5\%$ for all R as long as $T/a \geq 5$.[1]

In the following analysis, we perform numerical simulations at various β including $\beta = 6.600$. Since the width of the flux tube is considered to correspond to the length of APE smearing such that $a\sqrt{N_{\rm APE}}$ [19], we choose $N_{\rm APE}$ at other β so that $a\sqrt{N_{\rm APE}}$ be approximately constant.

---

[1] An alternative estimate of the ground state saturation is obtained through the excitation energy of a bosonic string: $\Delta E_n = \pi n/R$ ($n =, 1, 2, \ldots$) [21]. By taking $n = 2$, which corresponds to the excitation with the same symmetry (Σ_g^+) as the ground state [22], the excited state is suppressed at least by a factor $\exp(-(2\pi/R)T) = \exp(-\pi) \simeq 4\%$ for $T = 0.46$ fm and $R = 0.92$ fm, even if $C_{n=2}(R) \sim C_{n=0}(R)$.

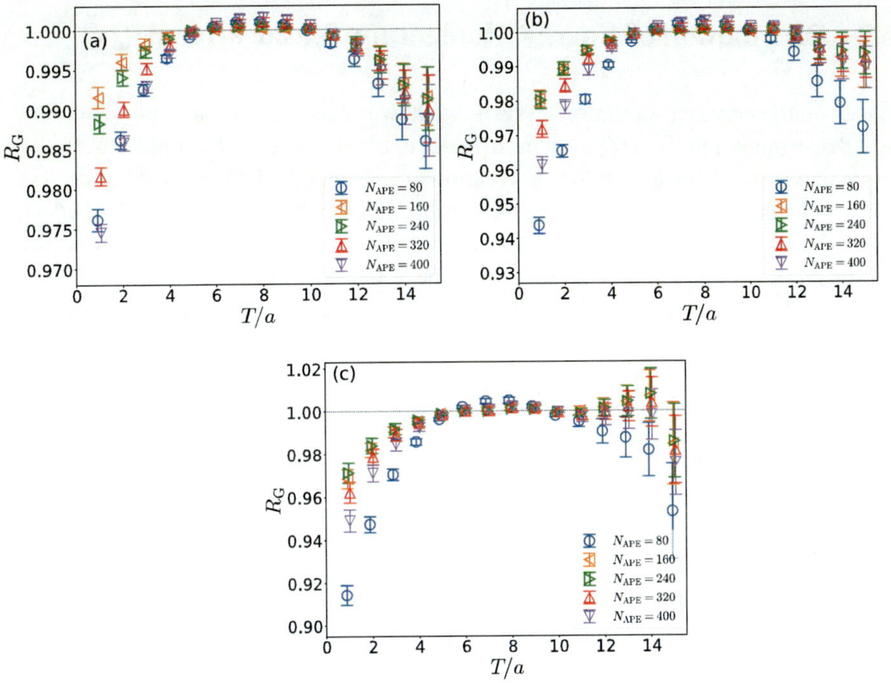

Fig. 6.4 T-dependence of R_G for various N_{APE}. From the left figure, (**a**) $R/a = 12$, (**b**) $R/a = 18$, and (**c**) $R/a = 24$

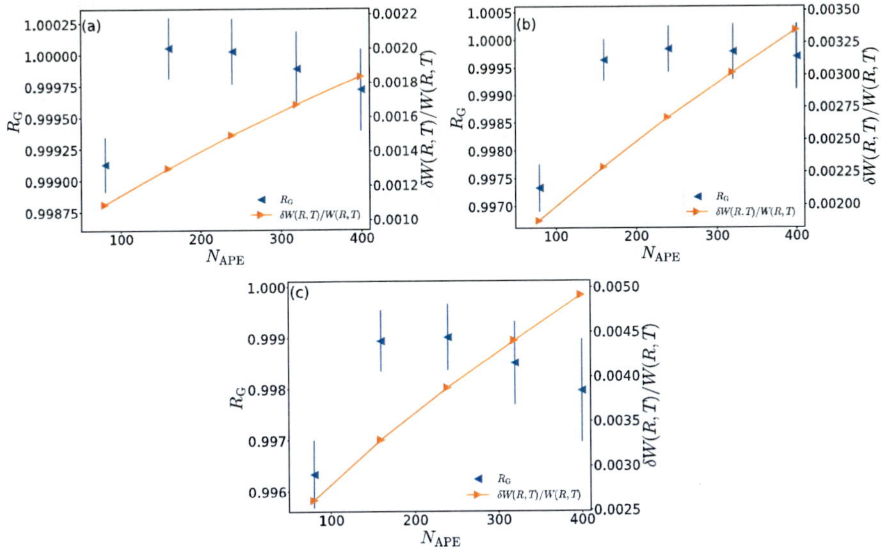

Fig. 6.5 N_{APE}-dependence of R_G and $\delta\langle W(R, T)\rangle/\langle W(R, T)\rangle$ at $T/a = 5$. From the left figure, (**a**) $R/a = 12$, (**b**) $R/a = 18$, and (**c**) $R/a = 24$

6.2 Distribution of Energy Momentum Tensor around $Q\bar{Q}$

Based on the previous auxiliary analysis, we now proceed to the measurement of the EMT distribution in the $Q\bar{Q}$ system. In order to obtain the distribution, we need two ingredients: the Wilson loop for the creation of $Q\bar{Q}$ and the EMT on the lattice. The former is discussed in the previous section in detail, while the latter is defined via the gradient flow according to the formula, Eq. (4.143) given in Chap. 4. Note that the matching coefficients $\alpha_U(t)$ and $\alpha_E(t)$ have one-loop accuracy as mentioned in Chap. 5. Here we discuss EMT in this inhomogeneous system and the coordinates useful for consideration.

6.2.1 *Correlation*

In quantum YM theory, obtaining EMT non-perturbatively around static $Q\bar{Q}$ on the lattice requires us the following correlation:

$$\langle \mathcal{T}_{\mu\nu}(t,x)\rangle_{Q\bar{Q}} = \lim_{T\to\infty} \frac{\langle \mathcal{T}_{\mu\nu}(t,x) W(R,T)\rangle_0}{\langle W(R,T)\rangle_0}, \tag{6.7}$$

where, in the rectangular Wilson loop, static charges reside at $\mathbf{R}_\pm = (0, 0, \pm R/2)$ and in the temporal interval $[-T/2, T/2]$. Here $\langle\cdot\rangle_0$ explicitly denotes taking the vacuum expectation value. The measurements of $\mathcal{T}_{\mu\nu}(t,x)$ in Eq. (4.143) for different flow times t are made at mid-temporal plane $x_\mu = (\mathbf{x}, x_4 = 0)$ and $W(R,T)$ is defined at $t = 0$.

In order to obtain the renormalized EMT distribution around $Q\bar{Q}$ from the lattice data, as mentioned in Chap. 4, we need to take the double limit [23],

$$\langle \mathcal{T}^{\mathrm{R}}_{\mu\nu}(x)\rangle_{Q\bar{Q}} = \lim_{t\to 0}\lim_{a\to 0} \langle \mathcal{T}_{\mu\nu}(t,x)\rangle^{\mathrm{lat}}_{Q\bar{Q}}. \tag{6.8}$$

In lattice simulations, we measure $\langle \mathcal{T}_{\mu\nu}(t,x)\rangle^{\mathrm{lat}}_{Q\bar{Q}}$ at finite t and a. The extrapolation to $(t,a) = (0,0)$ is made on the basis of the following parameterization,

$$\langle \mathcal{T}_{\mu\nu}(t,x)\rangle^{\mathrm{lat}}_{Q\bar{Q}} \approx \langle \mathcal{T}^{\mathrm{R}}_{\mu\nu}(x)\rangle_{Q\bar{Q}} + b_{\mu\nu}(t)a^2 + c_{\mu\nu}t, \tag{6.9}$$

where $b_{\mu\nu}(t)$ and $c_{\mu\nu}$ are contributions from lattice discretization effects and the dimension six operator, respectively. Note that the latter contribution results from the small flow time expansion and we employ the term up to the first order in t.

6.2.2 Cylindrical Coordinate

In our system under consideration where two static charges exist, since there exists a rotational symmetry around an axis connecting two charges, the cylindrical coordinate (r, θ, z) is useful, where $r = \sqrt{x^2 + y^2}$ and $\theta = \tan^{-1}(y/x)$ $(0 \leq \theta \leq 2\pi)$ with the Cartesian coordinates (Fig. 6.6).[2] The components of EMT in this coordinate system are given by

$$\mathcal{T}_{\gamma\gamma'} = (e_\gamma)_\mu \mathcal{T}_{\mu\nu}(e_{\gamma'})_\nu, \tag{6.10}$$

with $\gamma, \gamma' = r, \theta, z, 4$ and e_γ denotes a unit vector along the direction of the γ axis.

In particular, on the mid-plane between two charges, EMT is diagonalized as

$$\langle \mathcal{T}_{\gamma\gamma'}(t, x)\rangle^{\text{lat}}_{Q\bar{Q}} = \text{diag}(\langle \mathcal{T}_{44}(t, r)\rangle^{\text{lat}}_{Q\bar{Q}}, \langle \mathcal{T}_{rr}(t, r)\rangle^{\text{lat}}_{Q\bar{Q}}, \langle \mathcal{T}_{\theta\theta}(t, r)\rangle^{\text{lat}}_{Q\bar{Q}}, \langle \mathcal{T}_{zz}(t, r)\rangle^{\text{lat}}_{Q\bar{Q}}). \tag{6.11}$$

We remark that, in Maxwell theory, the magnitude of all these components degenerates as follows:

$$\mathcal{T}^{\text{Maxwell}}_{\gamma\gamma'}(\mathbf{x}) = \frac{1}{2}\text{diag}(-\mathbf{E}^2, \mathbf{E}^2, \mathbf{E}^2, -\mathbf{E}^2), \tag{6.12}$$

with the electric field $\mathbf{E}$.

6.3 Setup

The numerical simulations in the SU(3) YM theory were performed on the four-dimensional Euclidean lattice with the Wilson gauge action and the periodic boundary condition. Shown in Table 6.2 are five different inverse couplings $\beta = 6/g_0^2$ and corresponding lattice spacings a determined from the w_0-scale [23, 24]. The lattice size N_{size}^4 and the number of gauge configurations N_{conf} are also summarized in the table. Notice that the spatial volumes at both $\beta = 6.600$ and 6.819 are chosen to be the same in the physical unit and those on the coarser lattices than at $\beta = 6.600$ have larger volumes. N_{APE} is determined in the previous section with the same smearing parameter as in Ref. [19]. Also, to reduce the statistical noise, we adopt the standard multihit procedure by replacing each temporal link by its mean-field value [8, 20]. The numerical simulations were carried out on IBM System Blue Gene Solution at KEK.

---

[2] When one takes the ordinary principal value of the arctangent, the angle θ is to be restricted within $[-\pi/2, \pi/2]$. In order to precisely define θ with the Cartesian coordinates, one should employ either the transformation $\theta = \pi + \text{sgn}(y)\cos^{-1}(x/\sqrt{x^2 + y^2})$ or the so-called *atan2* function.

Table 6.2 First five rows are the simulation parameters on the lattice. Spatial size of the Wilson loop R is shown in the lattice and physical units. Temporal size of the Wilson loop is set to T unless otherwise stated

β	a (fm)	N_{size}^4	N_{conf}	N_{APE}	R/a			T/a
6.304	0.058	48^4	140	75	8	12	16	8
6.465	0.046	48^4	440	110	10	–	20	10
6.513	0.043	48^4	600	130	–	16	–	10
6.600	0.038	48^4	1,500	160	12	18	24	12
6.819	0.029	64^4	1,000	300	16	24	32	16
R (fm)					0.46	0.69	0.92	

Gauge configurations are generated by the same procedure as in Chap. 5 with the separation of 200 (100) sweeps on the 64^4 (48^4) lattice.[3] Statistical errors are estimated by the jackknife method with 20 jackknife bins at which the errors saturate. In the flow equation (4.48), the Wilson gauge action is used, while the clover-type representation is adopted for $G^a_{\mu\nu}(t, x)$ in $\mathcal{T}_{\mu\nu}(t, x)$. We remark that, in the EMT formula Eq. (4.143), we use the one-loop-order coefficients as $\alpha_U(t)$ and $\alpha_E(t)$.

We consider three $Q\bar{Q}$ distances ($R = 0.46, 0.69, 0.92$ fm), which are comparable to the typical scale of strong interaction. These values as well as the corresponding dimensionless distances R/a are summarized in Table 6.2. While the largest R is half the spatial lattice extent aN_{size} for the two finest lattice spacings, effects of the periodic boundary are known to be well suppressed even with this setting [8]. In fact, the $Q\bar{Q}$ potential we have calculated also shows almost no finite volume effect. The temporal length T is selected so that the physical value be approximately the same among various lattice spacings. Since, in the previous section, we have already confirmed that $T/a \geq 5$ at $\beta = 6.600$ is suitable to saturate the ground state in the Wilson loop, our present choice of T summarized in Table 6.2 is large enough to satisfy this condition. However, the smearing of the gauge field via the gradient flow has a possibility to affect the coupling to the ground state. We will discuss later the saturation of the correlation function between the EMT operator and the Wilson loop.

In order to avoid the artifact due to finite a and the over-smearing of the gradient flow [23, 26], we need to choose an appropriate window of t satisfying the condition $a/2 \lesssim \rho \lesssim L$. Here $\rho \equiv \sqrt{2t}$ is the flow radius, and $L \equiv \min(|\mathbf{x} - \mathbf{R}_+|, |\mathbf{x} - \mathbf{R}_-|, T/2)$ is the minimal distance between $x_\mu = (\mathbf{x}, 0)$ and the Wilson loop.

As a final remark of this section, we explicitly show the $Q\bar{Q}$ system we consider in Fig. 6.6. The origin resides at the mid-point between Q and $\bar{Q}$. We set the z axis

[3] Our simulation on fine lattices may suffer from the topological freezing [25]. However, from the analysis with the gauge configurations used for the scale setting in Ref. [23], it was confirmed that the dependence of $E(t, x)$ on the topological sector is less than 1% on 64^4 lattice at $\beta = 6.88$. With a reasonable assumption that the dependence of Eq. (6.7) on different topological sectors is of the same order, the topological freezing is likely to be less than the statistical errors in the present study.

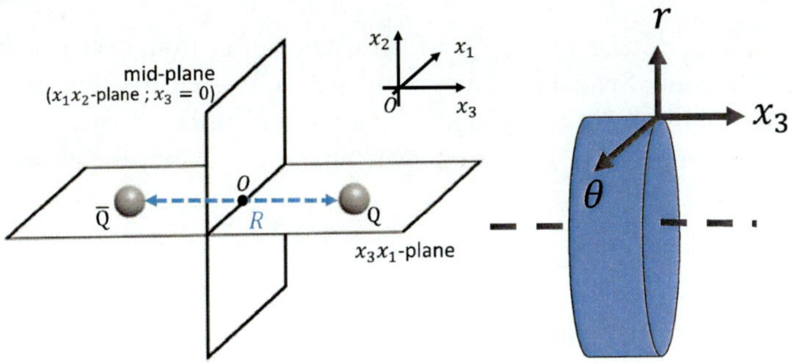

Fig. 6.6 (Left) $Q\bar{Q}$ system; (Right) cylindrical coordinate

to be the direction connecting two sources while xy plane is perpendicular to z-axis. In the following, we refer to yz plane and xy plane with $z = 0$ as 'source plane' and 'mid-plane,' respectively.

6.4 EMT Distribution on Source Plane

Before taking the double limit in Eq. (6.8), we illustrate a qualitative feature of the distribution of $\mathcal{T}_{ij}$ around $Q\bar{Q}$ in YM theory at fixed a and t by considering the case $a = 0.029$ fm with $R = 0.69$ fm, $T = 0.46$ fm, and $t/a^2 = 2.0$. Shown in Fig. 6.7 are the stress distribution around $Q\bar{Q}$ in the SU(3) YM theory and around two opposite charges in Maxwell theory. Figure 6.7(b) is the same plot as Fig. 2.3 as a comparison. Note that the other eigenvector is perpendicular to the yz plane. As is the case in Fig. 2.3, the eigenvector with negative (positive) eigenvalue is denoted by the red outward (blue inward) arrow with its length proportional to $\sqrt{|\lambda_k|}$.

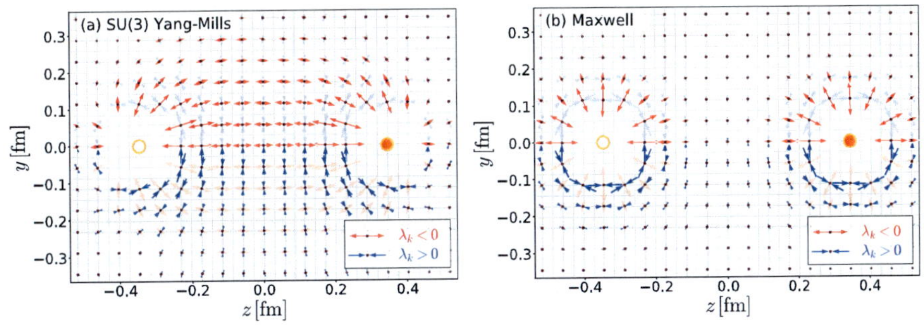

Fig. 6.7 (**a**) Distribution of the principal axes of $\mathcal{T}_{ij}$ for a $Q\bar{Q}$ system separated by $R = 0.69$ fm in the SU(3) Yang–Mills theory. The lattice spacing and the flow time are fixed to be $a = 0.029$ fm and $t/a^2 = 2.0$, respectively. (**b**) Distribution of the principal axes of the $\mathcal{T}_{ij}$ in classical electrodynamics between opposite charges. This figure is the same as Fig. 2.3. In both figures, the red (blue) arrows in the upper (lower) half plane are highlighted

The spatial regions near Q and $\bar{Q}$, which would suffer from over-smearing, are excluded in the figure. Spatial structure of the flux tube is clearly revealed through the stress tensor in Fig. 6.7(a) in a gauge invariant way. This is in contrast to the same plot of the principal axes of $\mathcal{T}_{ij}$ for opposite charges in classical electrodynamics given in Fig. 6.7(b).

6.5 EMT Distribution on Mid-Plane

For the more quantitative discussion, let us now turn to the mid-plane between the $Q\bar{Q}$ pair and extract the stress tensor distribution at $\mathbf{x} = (x, y, 0)$ by taking the double limit Eq. (6.8). Although we have briefly explained the double extrapolation, that is the continuum limit $a \to 0$ followed by $t \to 0$, to obtain the physical quantity in the four-dimensional Euclidean space in Chap. 4, we will show specific procedures with our actual data, such as a determination of a range employed for the extrapolation. Then, we show the resultant EMT distribution.

6.5.1 *Double Extrapolation*

We first focus on the mid-point $(0, 0, 0)$ between $Q\bar{Q}$. Note that all the data at different lattice spacings share the origin. In Fig. 6.8(a), we show an example of the T dependence of $\langle \mathcal{T}_{zz}(t, 0)\rangle^{\rm lat}_{Q\bar{Q}}$, which gives the largest eigenvalue on the mid-plane, for several values of t with $a = 0.029$ fm $(\equiv \bar{a})$. The figure indicates that significant T-dependence arises owing to over-smearing of the gradient flow for $t/\bar{a}^2 \sim 6$ (i.e., $\rho \sim \sqrt{12}\bar{a} = 0.10$ fm) already around $T = 16\bar{a} = 0.46$ fm. This is so for the same $t/\bar{a}^2$ in all other cases in Table 6.2. On the other hand, undersmearing of the gradient flow on the coarsest lattice becomes significant for $t/\bar{a}^2 = 2$ $(\rho = a = 0.058$ fm). Therefore, in the following analysis, we focus on the data in the interval $2 \le t/\bar{a}^2 \le 5$ $(2\bar{a} \le \rho \le \sqrt{10}\bar{a})$, which satisfies $a/2 \lesssim \rho \lesssim L$ with margin.

In Fig. 6.8(b), we show $\langle \mathcal{T}_{zz}(t, 0)\rangle^{\rm lat}_{Q\bar{Q}}$ as a function of the dimensionless ratio $a^2/\bar{a}^2$ by the open triangles for different values of t. The continuum limit $(a \to 0)$ with fixed t is taken by using these data together with the formula Eq. (6.9). The results are shown by the filled black squares with error bars.

In Fig. 6.8(c), the open symbols with error bars correspond to the original data for various values of a and $t/\bar{a}^2$. The result of the continuum limit in the interval $2 \le t/\bar{a}^2 \le 5$ is denoted by the black solid line with the shaded error band. The $t \to 0$ limit is carried out by using the values in the continuum limit according to Eq. (6.9) with $a = 0$. As a most conservative range for the extrapolation, we take $3 \le t/\bar{a}^2 \le 4$ (Range-1). Also, to estimate the systematic errors from the extrapolation, we consider two different ranges by changing the upper and lower limits: $2 \le t/\bar{a}^2 \le 4$ (Range-2) and $3 \le t/\bar{a}^2 \le 5$ (Range-3). The resulting values of $\langle \mathcal{T}^{\rm R}_{zz}(0)\rangle_{Q\bar{Q}}$ after the double

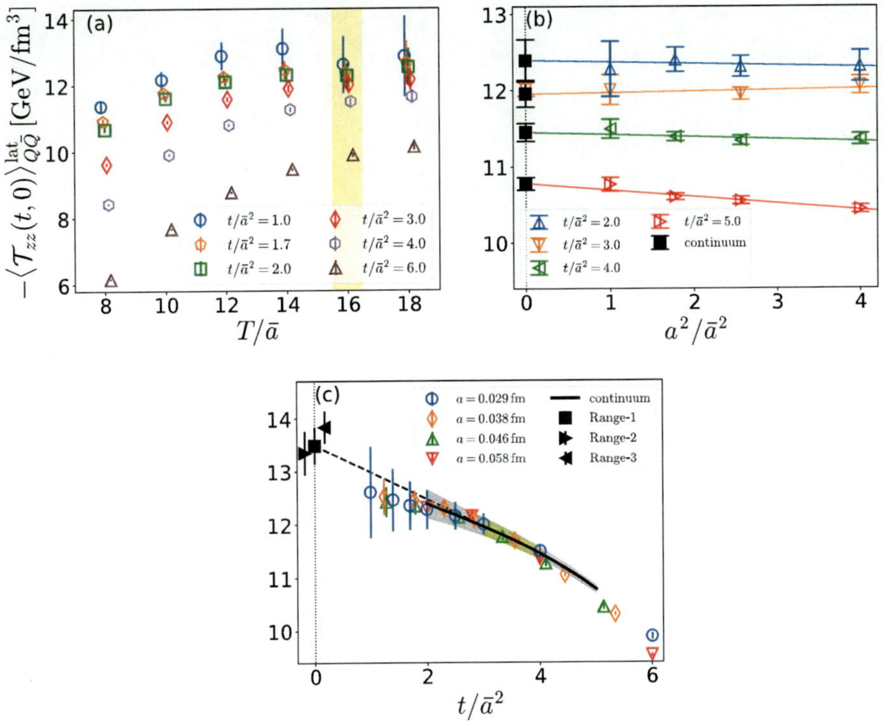

Fig. 6.8 zz component of the stress tensor at $\mathbf{x} = 0$, for various a and t with $R = 0.46$ fm. (**a**) Open symbols with the statistical errors are $\langle \mathcal{T}_{zz}(t,0)\rangle^{\mathrm{lat}}_{Q\bar{Q}}$ as a function of $T/\bar{a}$ with $a = \bar{a} = 0.029$ fm. To take the double limit, the data at $T/\bar{a} = 16$ indicated by the yellow band is used. (**b**) Open symbols with errors denote $\langle \mathcal{T}_{zz}(t,0)\rangle^{\mathrm{lat}}_{Q\bar{Q}}$ as a function of $a^2/\bar{a}^2$. The filled black symbols are the results of the $a \to 0$ limit for each t. (**c**) Open symbols with errors are $\langle \mathcal{T}_{zz}(t,0)\rangle^{\mathrm{lat}}_{Q\bar{Q}}$ as a function of $t/\bar{a}^2$ for different a. The solid line corresponds to the result of $a \to 0$ limit in the interval $2 \le t/\bar{a}^2 \le 5$ with shaded band being statistical error. The filled black symbols are the results of the $t \to 0$ extrapolation from three different ranges of t

limit are shown by the filled black symbols at $t = 0$. The dashed line corresponds to the extrapolation with Range-1.

The double limit for general $\mathbf{x} = (x, y, 0)$ on the mid-plane can be carried out essentially through the same procedure with a few extra steps. When we employ the cylindrical coordinate (r, θ, z) in the $Q\bar{Q}$ system, on the mid-plane we have a diagonalized EMT Eq. (6.11). We make the most of the cylindrical symmetry on the mid-plane to reduce the statistical error.

Next, we need data at the same r for different a to take the continuum limit. We consider the values of r at which the lattice data are available on the finest lattice. To obtain the data at these r on lattices with different a, we interpolate the lattice data $\langle \mathcal{T}_{\gamma\gamma}(t,r)\rangle^{\mathrm{lat}}_{Q\bar{Q}}$ with the commonly used functions to parameterize the transverse profile of the flux tube:

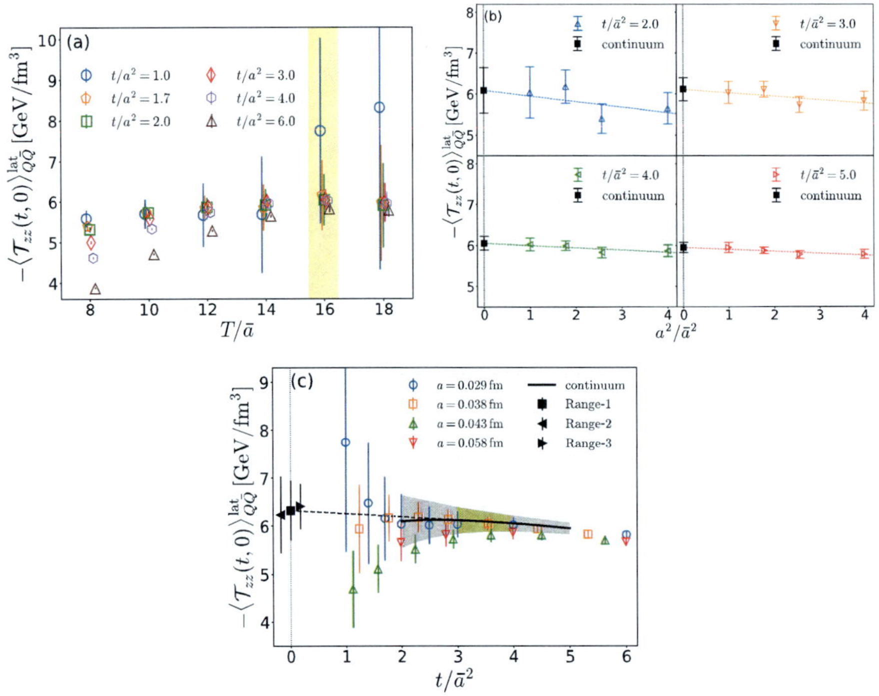

Fig. 6.9 Same analysis at $R = 0.69$ fm as Fig. 6.8

$$f_{\mathrm{Bessel}}(r) = A_0 K_0\left(\sqrt{Br^2 + C}\right), \tag{6.13}$$

with the 0th-order modified Bessel function $K_0(x)$ [27], and the function proposed in Ref. [14],

$$f_{\mathrm{exp}}(r) = (A_0 + A_1 r^2) e^{(-2\sqrt{r^2+B^2}+2B)/C}. \tag{6.14}$$

Once it is done, the $t \to 0$ limit is taken in the same way as explained above.

Shown in Figs. 6.9 and 6.10 are the corresponding plots at $R = 0.69, 0.92$ fm to Fig. 6.8, respectively. We also implement the same process as that at $R = 0.46$ fm although the statistical errors are larger.

6.5.2 *EMT Distribution on Mid-Plane*

In Fig. 6.11, we show the r dependence of the resulting EMT, the stress tensor as well as the energy density. From the figure, one finds several noticeable features:

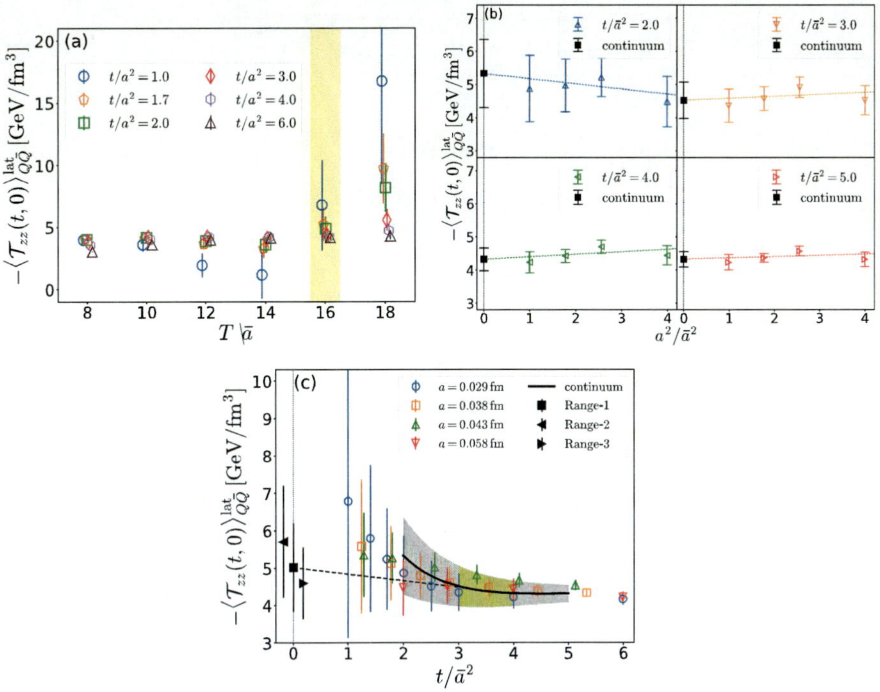

Fig. 6.10 Same analysis at $R = 0.92$ fm as Fig. 6.8

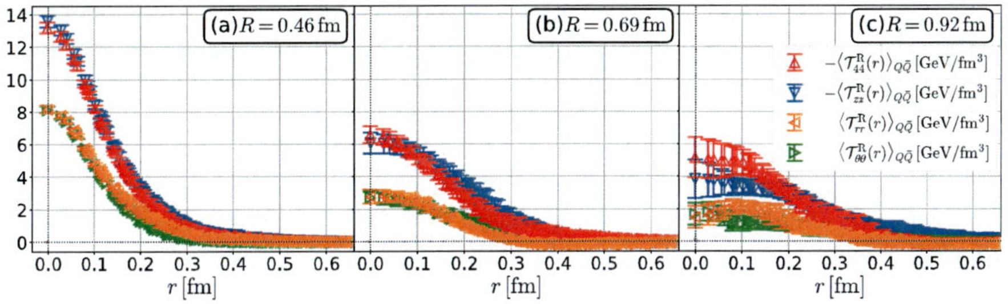

Fig. 6.11 EMT distribution on the mid-plane after the double limit $\langle \mathcal{T}^{\text{R}}_{\gamma\gamma'}(r)\rangle_{Q\bar{Q}}$ in the cylindrical coordinate system for three different values of the $Q\bar{Q}$ distance R

- Approximate degeneracy between temporal and longitudinal components is found for a wide range of r: $\langle \mathcal{T}^{\text{R}}_{44}(r)\rangle_{Q\bar{Q}} \simeq \langle \mathcal{T}^{\text{R}}_{zz}(r)\rangle_{Q\bar{Q}} < 0$. This feature is compatible with the leading-order prediction of the worldsheet theory of QCD string [12]. We also find $\langle \mathcal{T}^{\text{R}}_{rr}(r)\rangle_{Q\bar{Q}} \simeq \langle \mathcal{T}^{\text{R}}_{\theta\theta}(r)\rangle_{Q\bar{Q}} > 0$, which does not have simple interpretation except at $r = 0$.
- The scale symmetry broken in the YM vacuum (the trace anomaly) is partially restored inside the flux tube, which arises in the numerical results,

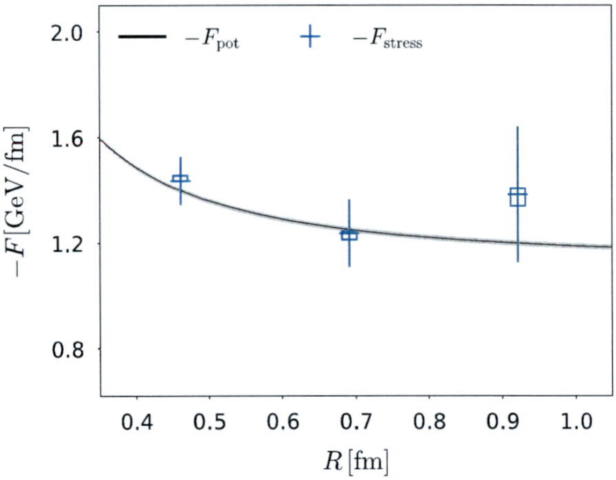

Fig. 6.12 R dependence of the $Q\bar{Q}$ forces, $-F_{\rm stress}$ and $-F_{\rm pot}$, obtained by the Wilson loop and the stress tensor, respectively. Error bars and rectangular boxes for the latter represent the statistical and systematic errors, respectively

$$\langle \mathcal{T}^{\rm R}_{\mu\mu}(r)\rangle_{Q\bar{Q}} = \langle \mathcal{T}^{\rm R}_{44}(r) + \mathcal{T}^{\rm R}_{zz}(r) + \mathcal{T}^{\rm R}_{rr}(r) + \mathcal{T}^{\rm R}_{\theta\theta}(r)\rangle_{Q\bar{Q}} < 0. \tag{6.15}$$

This is in sharp contrast to the case of classical electrodynamics; $\mathcal{T}_{44}(r) = \mathcal{T}_{zz}(r) = -\mathcal{T}_{rr}(r) = -\mathcal{T}_{\theta\theta}(r)$ and $\mathcal{T}_{\mu\mu}(r) = 0$ for all r.

- Each component of EMT at $r = 0$ decreases as R becomes larger, while the transverse radius of the flux tube, typically about 0.2 fm, seems to increase for large R [11, 14, 28, 29], although the statistics is not enough to discuss the radius quantitatively.

Finally, we perform a non-trivial consistency check of our analysis regarding the relation between the force acting on the charge located at $z > 0$ evaluated by the $Q\bar{Q}$ potential through

$$F_{\rm pot} = -\frac{dV(R)}{dR}, \tag{6.16}$$

and the force evaluated by the surface integral of the stress tensor surrounding the charge,

$$F_{\rm stress} = -\int \langle \mathcal{T}_{zj}(x)\rangle_{Q\bar{Q}}\, dS_j. \tag{6.17}$$

For $F_{\rm pot}$, we fit the numerical data of $V(R)$ in Fig. 6.2 obtained from the Wilson loop at $a = 0.038$ fm by the Cornell parameterization, Eq. (6.4). Note that $V(R)$ at this lattice spacing is shown to be already close to the continuum limit [4]. For $F_{\rm stress}$, we take the mid-plane for the surface integral:

$$F_{\mathrm{stress}} = 2\pi \int_0^\infty \langle \mathcal{T}_{zz}(r) \rangle_{Q\bar{Q}}\, r dr. \tag{6.18}$$

Here $\langle \mathcal{T}_{zz}(r) \rangle_{Q\bar{Q}}$ is obtained by fitting Fig. 6.11 with either $f_{\mathrm{Bessel}}(r)$ or $f_{\mathrm{exp}}(r)$. In Fig. 6.12, $-F_{\mathrm{pot}}$ and $-F_{\mathrm{stress}}$ thus obtained are shown by the solid line and the horizontal bars, respectively. $-F_{\mathrm{pot}}$ increases as $\mathcal{A}/R^2$ at short distance and approaches the string tension σ at large distance. For $-F_{\mathrm{stress}}$, we take into account not only the statistical error but also the systematic errors from the double limit and the fitting in terms of $f_{\mathrm{Bessel,exp}}(r)$. The agreement between the two quantities within the errors is a first numerical evidence that the 'action-at-a-distance' $Q\bar{Q}$ force can be described by the local properties of the stress tensor in YM theory.

References

1. Nambu Y (1974) Strings, monopoles and gauge fields. Phys Rev D 10:4262
2. t Hooft G, Gauge fields with unified weak, electromagnetic, and strong interactions. PRINT-75-0836 (UTRECHT)
3. Mandelstam S (1976) Vortices and quark confinement in nonabelian Gauge theories. Phys Rept 23:245
4. Bali GS (2001) QCD forces and heavy quark bound states. Phys Rept 343:1–136. arXiv:hep-ph/0001312 [hep-ph]
5. Kondo KI, Kato S, Shibata A, Shinohara T (2015) Quark confinement: dual superconductor picture based on a non-Abelian Stokes theorem and reformulations of Yang-Mills theory. Phys Rept 579:1–226. arXiv:1409.1599 [hep-th]
6. Greensite J (2011) An introduction to the confinement problem. Lect Notes Phys 821:1–211
7. Di Giacomo A, Maggiore M, Olejnik S (1990) Confinement and chromoelectric flux tubes in lattice QCD. Nucl Phys B 347:441–460
8. Bali GS, Schilling K, Schlichter C (1995) Observing long color flux tubes in SU(2) lattice gauge theory. Phys Rev D 51:5165–5198. arXiv:hep-lat/9409005 [hep-lat]
9. Michael C (1996) Lattice sum rules for the color fields. Phys Rev D 53:4102–4105. arXiv:hep-lat/9504016 [hep-lat]
10. Green AM, Michael C, Spencer PS (1997) The Structure of flux tubes in SU(2). Phys Rev D 55:1216–1225. arXiv:hep-lat/9610011 [hep-lat]
11. Gliozzi F, Pepe M, Wiese UJ (2010) The width of the confining string in Yang-Mills theory. Phys Rev Lett 104:232001. arXiv:1002.4888 [hep-lat]
12. Meyer HB (2010) The transverse structure of the QCD string. Phys Rev D 82:106001. arXiv:1008.1178 [hep-lat]
13. Cea P, Cosmai L, Papa A (2012) Chromoelectric flux tubes and coherence length in QCD. Phys Rev D 86:054501. arXiv:1208.1362 [hep-lat]
14. Cardoso N, Cardoso M, Bicudo P (2013) Inside the SU(3) quark-antiquark QCD flux tube: screening versus quantum widening. Phys Rev D 88:054504. arXiv:1302.3633 [hep-lat]
15. Narayanan R, Neuberger H (2006) Infinite N phase transitions in continuum wilson loop operators. JHEP 0603:064 [hep-th/0601210]
16. Lüscher M (2010) Properties and uses of the wilson flow in lattice QCD. JHEP 1008:071. arXiv:1006.4518 [hep-lat]
17. Lüscher M, Weisz P (2011) Perturbative analysis of the gradient flow in non-abelian gauge theories. JHEP 1102:051. arXiv:1101.0963 [hep-th]
18. Albanese M et al (1987) [APE Collaboration], Glueball masses and string tension in lattice QCD. Phys Lett B 192:163

19. Takahashi TT, Suganuma H, Nemoto Y, Matsufuru H (2002) Detailed analysis of the three quark potential in SU(3) lattice QCD. Phys Rev D 65:114509. arXiv:hep-lat/0204011 [hep-lat]
20. Parisi G, Petronzio R, Rapuano F (1983) A Measurement of the String Tension Near the Continuum Limit. Phys Lett 128B:418
21. Lüscher M, Weisz P (2002) Quark confinement and the bosonic string. JHEP 07:049. arXiv:hep-lat/0207003 [hep-lat]
22. Juge KJ, Kuti J, Morningstar C (2003) Fine structure of the QCD string spectrum. Phys Rev Lett 90:161601. arXiv:hep-lat/0207004 [hep-lat]
23. Kitazawa M, Iritani T, Asakawa M, Hatsuda T, Suzuki H (2016) Equation of state for SU(3) Gauge theory via the energy-momentum tensor under gradient flow. Phys Rev D 94(11):114512. arXiv:1610.07810 [hep-lat]
24. Borsanyi S, Durr S, Fodor Z, Hoelbling C, Katz SD, Krieg S, Kurth T, Lellouch L, Lippert T, McNeile C et al (2012) High-precision scale setting in lattice QCD. JHEP 09:010 [arXiv:1203.4469 [hep-lat]]
25. Endres MG (2016) New simulation strategies for lattice gauge theory. PoS LATTICE2016:014. arXiv:1612.01609 [hep-lat]
26. Kitazawa M, Iritani T, Asakawa M, Hatsuda T (2017) Correlations of the energy-momentum tensor via gradient flow in SU(3) Yang-Mills theory at finite temperature. Phys Rev D 96(11):111502. arXiv:1708.01415 [hep-lat]
27. Clem JR (1975) Simple model for the vortex core in a type II superconductor. J Low Temp Phys 18:427
28. Lüscher M, Munster G, Weisz P (1981) How thick are chromoelectric flux tubes? Nucl Phys B 180:1–12
29. Gliozzi F, Pepe M, Wiese UJ (2010) The width of the color flux tube at 2-loop order. JHEP 11:053. arXiv:1006.2252 [hep-lat]

Chapter 7
Distribution of Energy Momentum Tensor around Magnetic Vortex in Abelian-Higgs Model

In Chap. 6, the distribution of the energy momentum tensor (EMT) in static quark and an anti-quark ($Q\bar{Q}$) systems in vacuum of the SU(3) Yang–Mills (YM) theory has been measured via the numerical simulation of lattice gauge theory. The analysis of EMT on the lattice is realized with the EMT operator [1–6] constructed via the gradient flow [7–9]. Through the analysis of the principal directions and eigenvalues of the stress tensor, the formation of the flux tube is revealed in terms of the gauge-invariant observable. We also find remarkable non-Abelian features by studying the transverse structure quantitatively. In addition, as a consistency check of our analysis, we confirm that the linear confining behavior of the $Q\bar{Q}$ potential at long distances can be reproduced by the surface integral of the stress tensor.

In this chapter, motivated by the numerical results investigated in Chap. 6, we study the distribution of EMT in the $Q\bar{Q}$ system using conservation laws and the Abelian-Higgs (AH) model. First, we show that the momentum conservation results in the indispensability of boundaries of the flux tube in order to describe the lattice results.

Then, we proceed to the study of the EMT distribution around the magnetic vortex with and without boundaries via the AH model. The AH model is a relativistically generalized version of the Ginzburg–Landau (GL) model for superconductivity. According to the dual superconductor picture [10–13], the dual of the AH model is regarded as a phenomenological model of low-energy QCD; attempts to derive the dual AH model from QCD have been discussed in the literatures [14–18] based on the Abelian dominance [19–21] and the monopole condensation [22–24]. In this picture, magnetic monopoles and the magnetic vortex between the monopoles in the AH model correspond to the color charges and the flux tube in YM theory, respectively. Based on this picture, the vortex solution in the AH model has been compared with numerical results on the flux tube in YM theory [25–32]. In these studies, however, the action density and/or the field strength of the color gauge field in a specific gauge have been used for observables to make a comparison.

R. Yanagihara, *Distribution of Energy Momentum Tensor around Static Charges in Lattice Simulations and an Effective Model*, Springer Theses,
https://doi.org/10.1007/978-981-16-6234-8_7

In contrast, we calculate the spatial distribution of EMT around the magnetic vortex in the AH model and compare it with the lattice result. We demonstrate that the EMT distribution around the infinitely long magnetic vortex is qualitatively inconsistent with the lattice result, as anticipated from the momentum conservation. We then analyze the magnetic vortex with a finite length and show that a wide parameter range of the AH model with a standard potential cannot reproduce the lattice result simultaneously.

The organization of this chapter is as follows. In Sect. 7.1, we consider the momentum conservation in the cylindrical coordinates. We introduce the AH model in Sect. 7.2 and then, in Sect. 7.3, discuss the magnetic vortex solution in this model with and without boundaries. In Sect. 7.4, we give the numerical results on the EMT distribution around the magnetic vortex.

Note that, throughout this chapter, we consider (3 + 1)-dimensional Minkowski flat space–time with the metric $g_{\mu\nu}=g^{\mu\nu}=\mathrm{diag}(1,-1,-1,-1)$ (μ,ν=0, 1, 2, 3).

7.1 Stress Tensor and Momentum Conservation in Cylindrical Coordinate

Before entering the analyses with the AH model, we consider the general properties of the EMT distribution which does not depend on specific models. We focus on the conservation law, $\partial_\mu \mathcal{T}^{\mu\nu}=0$, which reads

$$\partial_i \mathcal{T}^{ij}=0, \tag{7.1}$$

in a static system. By taking a volume integral on both sides of Eq. (7.1) on a volume without external charges, we confirm the equilibrium of the force.

Since the $Q\bar{Q}$ system has the cylindrical symmetry, the cylindrical coordinate (r,θ,z) is useful, where $r=\sqrt{x^2+y^2}$ and $\theta=\tan^{-1}(y/x)$ with $0\le\theta\le 2\pi$.[1] When we employ the components of EMT in this coordinate system given by

$$\mathcal{T}_{\gamma\gamma'}(r,z)=(e_\gamma)_\mu \mathcal{T}^{\mu\nu}(e_{\gamma'})_\nu, \tag{7.2}$$

with $\gamma,\gamma'=0,r,\theta,z$ and e_γ denotes a unit vector along the direction of the γ-axis in the Minkowski space, the conservation law, Eq. (7.1) is explicitly written down as

$$\frac{1}{r}\partial_r(r\mathcal{T}_{rr})-\frac{\mathcal{T}_{\theta\theta}}{r}+\partial_z\mathcal{T}_{rz}=0, \tag{7.3}$$

$$\partial_\theta \mathcal{T}_{\theta\theta}=0, \tag{7.4}$$

$$\frac{1}{r}\partial_r(r\mathcal{T}_{rz})+\partial_z\mathcal{T}_{zz}=0. \tag{7.5}$$

[1] One should refer to the remark on the definition of the angle θ in Chap. 6.

Each equation corresponds to the equilibrium of the force in an infinitesimal volume element along the r, θ, z directions, respectively.

When a system is translationally symmetric along the z direction in addition to the rotational symmetry, the components of EMT are given by functions of r as $\mathcal{T}_{\gamma\gamma'} = \mathcal{T}_{\gamma\gamma'}(r)$, and z derivatives in Eqs. (7.3)–(7.5) vanish. Only a non-trivial constraint from the momentum conservation is then obtained from Eq. (7.3) as

$$\partial_r (r\mathcal{T}_{rr}(r)) = \mathcal{T}_{\theta\theta}(r). \tag{7.6}$$

By rewriting Eq. (7.6) as $r\partial_r(\mathcal{T}_{rr}(r)) = \mathcal{T}_{\theta\theta}(r) - \mathcal{T}_{rr}(r)$, it is easy to show that $\mathcal{T}_{rr}(r)$ and $\mathcal{T}_{\theta\theta}(r)$ behave differently as functions of r except for the case $\mathcal{T}_{rr}(r) = \mathcal{T}_{\theta\theta}(r) = 0$ provided that $\mathcal{T}_{rr}(r) \to 0$ in the $r \to \infty$ limit. Moreover, by integrating out both sides of Eq. (7.6) and assuming $r\mathcal{T}_{rr}(r)|_{r\to\infty} = r\mathcal{T}_{rr}(r)|_{r\to 0} = 0$ one obtains

$$\int_0^\infty dr\, \mathcal{T}_{\theta\theta}(r) = 0, \tag{7.7}$$

which shows that $\mathcal{T}_{\theta\theta}(r)$ as a function of r must change sign at least once so that its integral vanishes.

Now, let us inspect the components of EMT on the mid-plane of the flux tube in the SU(3) YM theory obtained on the lattice shown in Fig. 6.11. The figure shows that $\mathcal{T}_{rr}(r)$ and $\mathcal{T}_{\theta\theta}(r)$ are degenerated within statistical errors for all $Q\bar{Q}$ distances. Moreover, the result suggests $\int_0^\infty dr\mathcal{T}_{\theta\theta}(r) > 0$. These properties do not agree with Eqs. (7.6) and (7.7) obtained by assuming the translational invariance. Therefore, the result in Fig. 6.11 shows that the assumption of translational invariance is not applicable to the flux tube in the SU(3) YM theory even on the mid-plane with $R = 0.92$ fm.

7.2 Abelian-Higgs Model

Confinement in QCD as well as the SU(3) YM theory is one of the non-perturbative phenomena in the low-energy region. In order to understand the confinement mechanism, Nambu [10], 't Hooft [11], and Mandelstam [12] proposed that the dual picture to the well-known superconductivity can explain the linear confining potential; hence, the color electric field is squeezed into one-dimensional tube structure, that is, the flux tube.

The AH model is a relativistic extension of the GL theory in the superconductivity [33] and the dual AH model is regarded as a low-energy effective model for QCD to investigate the confinement phenomena. As the dual picture of the magnetic vortex formed in the condensate of the Cooper pair in the case of the type-II superconductor, the electric field is excluded and is hence squeezed in the vacuum with the magnetic monopole condensate. When we regard the electric charge as the color charge, we can explain the fact that the color electric field is squeezed one-dimensionally in vacuum.

In the following, in order to investigate the structure of the flux tube, we analyze the EMT distribution around the magnetic vortex in the AH model. Note that we employ the AH model instead of the dual AH model since the essential idea is the same except for whether we employ the electric vortex between electric charges or the magnetic vortex between the magnetic monopoles.

7.2.1 Model

Now, we employ the AH model and investigate the EMT distribution around the magnetic vortex with and without boundaries. Our starting point is the AH Lagrangian in four-dimensional Minkowski space:

$$\begin{aligned}\mathcal{L}_{\text{AH}} &= -\frac{1}{4g^2}(\partial_\mu A_\nu(x) - \partial_\nu A_\mu(x))^2 + |(\partial_\mu + iA_\mu(x))\chi(x)|^2 - \lambda(|\chi(x)|^2 - v^2)^2 \\ &= -\frac{1}{4g^2}F_{\mu\nu}^2(x) + |D_\mu\chi(x)|^2 - V(|\chi(x)|),\end{aligned} \quad (7.8)$$

where $F_{\mu\nu}(x) = \partial_\mu A_\nu(x) - \partial_\nu A_\mu(x)$ and $D_\mu = \partial_\mu + iA_\mu(x)$ are the field strength tensor and the covariant derivative, respectively, with the Abelian gauge field $A_\mu(x)$ and the complex scalar field $\chi(x)$. The last term $V(|\chi(x)|) = \lambda(|\chi(x)|^2 - v^2)^2$ represents the Higgs potential which induces the scalar condensation. The AH model has three parameters, g, λ, and v. g is the gauge coupling constant, and v is the vacuum expectation value of $\chi(x)$. The EMT in the AH model is obtained as a Noether current of the translational symmetry as [34, 35]

$$\mathcal{T}_{\mu\nu}(x) = (D_\mu\chi(x))^*(D_\nu\chi(x)) + (D_\nu\chi(x))^*(D_\mu\chi(x)) - \frac{1}{g^2}g^{\rho\sigma}F_{\mu\rho}(x)F_{\nu\sigma}(x) - g_{\mu\nu}\mathcal{L}. \quad (7.9)$$

The AH model has two characteristic length scales: the correlation lengths of the scalar and gauge fields [36],

$$\xi_\chi = \frac{1}{2\sqrt{\lambda}v}, \quad \xi_A = \frac{1}{\sqrt{2}gv}, \quad (7.10)$$

respectively. The ratio of these two scales characterizes the GL parameter κ,

$$\kappa = \frac{\xi_\chi}{\sqrt{2}\xi_A} = \frac{\sqrt{\lambda}}{g}. \quad (7.11)$$

In the context of superconductivity, the GL parameter classifies the type of superconductor into type-I ($\kappa < 1/\sqrt{2}$) or type-II ($\kappa > 1/\sqrt{2}$). In the type-I superconductor, the interaction between vortices is attractive so that a system of two and more individual vortices is unstable for the formation of a single vortex with a multiple winding number. In contrast, the interaction between vortices is repulsive in the type-II superconductor [36]. The boundary value $\kappa = 1/\sqrt{2}$ is called the Bogomol'nyi bound [34].

The AH Lagrangian is invariant under the U(1) gauge transformation defined by

$$\chi(x) \to \chi(x)e^{i\alpha(x)}, \quad \chi^*(x) \to \chi^*(x)e^{-i\alpha(x)}, \quad A_\mu(x) \to A_\mu(x) - \partial_\mu\alpha(x), \tag{7.12}$$

with a gauge function $\alpha(x)$. The complex scalar field $\chi(x)$ is written as

$$\chi(x) = \phi(x)e^{if(x)}, \tag{7.13}$$

with a real scalar field $\phi(x)$ and a phase function $f(x)$. In the unitary gauge defined by $\alpha(x) = -f(x)$, we have the gauge-fixed AH Lagrangian,

$$\mathcal{L}_{\mathrm{AH}} = -\frac{1}{4g^2}(\partial_\mu A'_\nu(x) - \partial_\nu A'_\mu(x))^2 + |(\partial_\mu + iA'_\mu(x))\phi(x)|^2 - \lambda(\phi(x)^2 - v^2)^2, \tag{7.14}$$

with $A'_\mu(x) = A_\mu(x) + \partial_\mu f(x)$.

7.2.2 Energy Momentum Tensor in Cylindrical Coordinate System

In this study, we investigate the spatial distribution of EMT around the magnetic vortex with and without boundaries. As these systems possess rotational symmetry around the vortex, we employ the cylindrical coordinate system. For a vortex with length R, we suppose that the boundaries are at $(r, z) = (0, \pm R/2)$. From the rotational symmetry, the scalar field is given by a function of r and z as

$$\phi = \phi(r, z), \tag{7.15}$$

and it is possible to represent the gauge field as

$$\mathbf{A}(r, z) = \frac{\tilde{A}(r, z)}{r}\mathbf{e}_\theta \tag{7.16}$$

with a scalar function $\tilde{A}(r, z)$ and the three-dimensional unit vector $\mathbf{e}_\theta$ in the direction of the θ-axis.

The non-vanishing components of EMT in the cylindrical coordinate system, Eq. (7.2), are calculated to be

$$\mathcal{T}_{00}(r,z) = \frac{1}{2g^2r^2}\Big((\partial_r\tilde{A})^2 + (\partial_z\tilde{A})^2\Big) + \Big((\partial_r\phi)^2 + (\partial_z\phi)^2\Big) + \frac{\phi^2\tilde{A}^2}{r^2} + \lambda(\phi^2 - v^2)^2, \tag{7.17}$$

$$\mathcal{T}_{rr}(r,z) = \frac{1}{2g^2r^2}\Big((\partial_r\tilde{A})^2 - (\partial_z\tilde{A})^2\Big) + \Big((\partial_r\phi)^2 - (\partial_z\phi)^2\Big) - \frac{\phi^2\tilde{A}^2}{r^2} - \lambda(\phi^2 - v^2)^2, \tag{7.18}$$

$$\mathcal{T}_{\theta\theta}(r,z) = \frac{1}{2g^2r^2}\Big((\partial_r\tilde{A})^2 + (\partial_z\tilde{A})^2\Big) - \Big((\partial_r\phi)^2 + (\partial_z\phi)^2\Big) + \frac{\phi^2\tilde{A}^2}{r^2} - \lambda(\phi^2 - v^2)^2, \tag{7.19}$$

$$\mathcal{T}_{zz}(r,z) = \frac{-1}{2g^2r^2}\Big((\partial_r\tilde{A})^2 - (\partial_z\tilde{A})^2\Big) - \Big((\partial_r\phi)^2 - (\partial_z\phi)^2\Big) - \frac{\phi^2\tilde{A}^2}{r^2} - \lambda(\phi^2 - v^2)^2, \tag{7.20}$$

$$\mathcal{T}_{rz}(r,z) = \frac{1}{g^2r^2}(\partial_r\tilde{A})(\partial_z\tilde{A}) + 2(\partial_r\phi)(\partial_z\phi). \tag{7.21}$$

On the mid-plane at $z = 0$, $\mathcal{T}_{rz}(r, z = 0)$ vanishes from the symmetric reason and EMT is diagonalized as

$$\mathcal{T}_{\gamma\gamma'}(r) = \mathrm{diag}(\mathcal{T}_{00}(r), \mathcal{T}_{rr}(r), \mathcal{T}_{\theta\theta}(r), \mathcal{T}_{zz}(r)), \tag{7.22}$$

where the argument $z = 0$ in $\mathcal{T}_{\gamma\gamma'}(r, z)$ is abbreviated for notational simplicity. Each component is explicitly given by

$$\mathcal{T}_{00}(r) = -\mathcal{T}_{zz}(r) = \frac{1}{2g^2r^2}(\partial_r\tilde{A})^2 + (\partial_r\phi)^2 + \frac{\phi^2\tilde{A}^2}{r^2} + \lambda(\phi^2 - v^2)^2, \tag{7.23}$$

$$\mathcal{T}_{rr}(r) = \frac{1}{2g^2r^2}(\partial_r\tilde{A})^2 + (\partial_r\phi)^2 - \frac{\phi^2\tilde{A}^2}{r^2} - \lambda(\phi^2 - v^2)^2, \tag{7.24}$$

$$\mathcal{T}_{\theta\theta}(r) = \frac{1}{2g^2r^2}(\partial_r\tilde{A})^2 - (\partial_r\phi)^2 + \frac{\phi^2\tilde{A}^2}{r^2} - \lambda(\phi^2 - v^2)^2, \tag{7.25}$$

where we have used the fact that terms including ∂_z vanish on the mid-plane.

Equations (7.23)–(7.25) tell us several notable features. First, the absolute values of $\mathcal{T}_{00}(r)$ and $\mathcal{T}_{zz}(r)$ on the mid-plane always degenerate in the AH model. Second, the first term of Eqs. (7.23)–(7.25) corresponds to the Maxwell stress. On the mid-plane, the direction of the magnetic field is along the z-axis. As a result, as in Eq. (6.12) this term gives a negative (positive) contribution to $\mathcal{T}_{zz}(r)$ ($\mathcal{T}_{rr}(r)$ and $\mathcal{T}_{\theta\theta}(r)$). Third, the last term is a contribution from the Higgs potential. The contribution of this term is negative for all spatial components $\mathcal{T}_{zz}(r)$, $\mathcal{T}_{rr}(r)$, and $\mathcal{T}_{\theta\theta}(r)$, while the contribution to $\mathcal{T}_{00}(r)$ is positive. These contributions are understood as the negative pressure owing to the instability of a state having a deviation of ϕ from its vacuum expectation value v. Fourth, because $\mathcal{T}_{rr}(r) = \mathcal{T}_{\theta\theta}(r)$ at $r = 0$, one obtains $(\partial_r\phi)^2 = \phi^2\tilde{A}^2/r^2$ in the $r \to 0$ limit. As a result, the signs of $\mathcal{T}_{rr}(r)$ and $\mathcal{T}_{\theta\theta}(r)$ at $r = 0$ are determined by the interplay between the first and fourth terms, i.e., contributions from the gauge field and the Higgs potential, respectively.

7.3 Magnetic Vortex

In this section, we discuss the magnetic monopoles and the classical solution of the magnetic vortex between the monopoles in the AH model with and without boundaries.

7.3.1 *Magnetic Vortex with Finite Length*

Let us first consider a magnetic vortex with a finite length R between two magnetic monopoles with opposite charges. As shown by Dirac [37], the Maxwell theory can have monopoles with quantized charges. These monopoles are associated with a singularity of the gauge field called the Dirac string. When two monopoles with opposite charges are at $(r, z) = (0, \pm R/2)$, the Dirac string can be located on the z-axis between two monopoles. The gauge field in this case is given by

$$\mathbf{A}^D(r,z) = -\frac{n}{2r}\left(\frac{z+R/2}{\sqrt{r^2+(z+R/2)^2}} - \frac{z-R/2}{\sqrt{r^2+(z-R/2)^2}}\right)\mathbf{e}_\theta = \frac{\tilde{A}^D(r,z)}{r}\mathbf{e}_\theta, \tag{7.26}$$

where the winding number n corresponds to the charge of a monopole at $z = R/2$ in the unit of $2\pi/g$. One easily finds that the magnetic field $\mathbf{B}^D = \nabla \times \mathbf{A}^D$ is given by the superposition of the Coulombic fields with the charges $\pm 2n\pi/g$ at $(r, z) = (0, \pm R/2)$. The magnetic field at the origin thus is given by

$$|\mathbf{B}(0)| = 2\frac{2n\pi}{g}\frac{1}{4\pi(R/2)^2} = \frac{4n}{gR^2}. \tag{7.27}$$

Then, one obtains

$$\mathcal{T}_{00}(0) = \mathcal{T}_{rr}(0) = \mathcal{T}_{\theta\theta}(0) = \frac{8n^2}{g^2R^4}, \quad \mathcal{T}_{zz}(0) = -\frac{8n^2}{g^2R^4}, \tag{7.28}$$

on the mid-plane from the EMT formula in the Maxwell theory. In what follows, we consider the vortex with $n = 1$.

The Dirac monopoles can also be introduced in the AH model. As the gauge field in this case has the same singularity as Eq. (7.26), when monopoles are located at $(r, z) = (0, \pm R/2)$, it is convenient to denote the gauge field as [17, 18, 30]

$$\mathbf{A}(r, z) = \mathbf{A}^D(r, z) + \mathbf{a}(r, z), \tag{7.29}$$

$$\mathbf{a} = a(r, z)\mathbf{e}_\theta = \frac{\tilde{a}(r, z)}{r}\mathbf{e}_\theta. \tag{7.30}$$

Here, $\tilde{a}(r, z)$ does not have a singularity and represents the deviation of $\mathbf{A}(r, z)$ from Eq. (7.26). The classical solution of the magnetic vortex is then obtained by minimizing the total energy,

$$E = \int d^3x\, \mathcal{T}_{00}(r, z) = 2\pi \int_0^\infty r dr \int_{-\infty}^\infty dz\, \mathcal{T}_{00}(r, z), \tag{7.31}$$

as a functional of $\tilde{a}(r, z)$ and $\phi(r, z)$ with the boundary conditions,

$$\tilde{a}(r, z) \to 0 \qquad \text{for } r \to 0,\ \forall z, \tag{7.32}$$

$$\phi(r, z) \to 0 \qquad \text{for } r \to 0,\ -R/2 \le z \le R/2, \tag{7.33}$$

$$\tilde{a}(r, z) \to -\tilde{A}^D(r, z),\ \phi(r, z) \to v \qquad \text{for } r,\ z \to \infty. \tag{7.34}$$

The condition, Eq. (7.34), ensures that the total energy of this system is finite.

We note that $\tilde{a}(r, z)$ and $\phi(r, z)$ hardly change within the length ξ_A and ξ_χ. Therefore, when the condition $R \ll \xi_A, \xi_\chi$ is satisfied, they remain Eqs. (7.32) and (7.33) and take values almost equal to zero around the monopoles. In this case, the EMT around the monopoles should be dominated by the contribution from the gauge field $\mathbf{A}^D(r, z)$.

7.3.2 *Infinitely Long Vortex*

Next, we consider a magnetic vortex with an infinite length. This solution is obtained by taking the $R \to \infty$ limit in the above argument. In this limit, the system has a translational symmetry along the z direction, and ϕ and $\tilde{a}$ are given by functions of only r as $\phi = \phi(r)$ and $\tilde{a} = \tilde{a}(r)$. By taking the $R \to \infty$ limit of Eq. (7.26) with $n = 1$, one obtains

$$\tilde{A}^D(r, z) = -1. \tag{7.35}$$

The vortex solution is obtained by minimizing the energy per unit length,

$$\int_0^\infty (2\pi r)dr\, \mathcal{T}_{00}(r), \tag{7.36}$$

with respect to $\phi(r)$ and $\tilde{a}(r)$ with the boundary conditions,

$$\tilde{a}(r) \to 0,\ \phi(r) \to 0 \qquad \text{for } r \to 0, \tag{7.37}$$

$$\tilde{a}(r) \to 1,\ \phi(r) \to v \qquad \text{for } r \to \infty. \tag{7.38}$$

It is convenient to introduce dimensionless variables,

$$\rho = rgv, \quad P(\rho) = \frac{\phi(\rho/gv)}{v}, \quad Q(\rho) = \tilde{A}(\rho/gv). \tag{7.39}$$

Using these variables, the components of the dimensionless EMT,

$$\hat{\mathcal{T}}_{\gamma\gamma'}(\rho) = \frac{\xi_A^2}{v^2}\mathcal{T}_{\gamma\gamma'}\left(\frac{\rho}{gv}\right), \tag{7.40}$$

are given by

$$\hat{\mathcal{T}}_{00}(\rho) = -\hat{\mathcal{T}}_{zz}(\rho) = \hat{\mathcal{T}}_{(1)}(\rho) + \hat{\mathcal{T}}_{(2)}(\rho) + \hat{\mathcal{T}}_{(3)}(\rho) + \hat{\mathcal{T}}_{(4)}(\rho), \tag{7.41}$$

$$\hat{\mathcal{T}}_{rr}(\rho) = \hat{\mathcal{T}}_{(1)}(\rho) + \hat{\mathcal{T}}_{(2)}(\rho) - \hat{\mathcal{T}}_{(3)}(\rho) - \hat{\mathcal{T}}_{(4)}(\rho), \tag{7.42}$$

$$\hat{\mathcal{T}}_{\theta\theta}(\rho) = \hat{\mathcal{T}}_{(1)}(\rho) - \hat{\mathcal{T}}_{(2)}(\rho) + \hat{\mathcal{T}}_{(3)}(\rho) - \hat{\mathcal{T}}_{(4)}(\rho), \tag{7.43}$$

with

$$\hat{\mathcal{T}}_{(1)}(\rho) = \frac{(\partial_\rho Q)^2}{4\rho^2}, \quad \hat{\mathcal{T}}_{(2)}(\rho) = \frac{(\partial_\rho P)^2}{2}, \quad \hat{\mathcal{T}}_{(3)}(\rho) = \frac{P^2Q^2}{2\rho^2}, \quad \hat{\mathcal{T}}_{(4)}(\rho) = \frac{\kappa^2(P^2-1)^2}{2}. \tag{7.44}$$

The energy per unit length, Eq. (7.36), is represented as

$$\int_0^\infty (2\pi r)dr\, \mathcal{T}_{00}(r) = 2\pi v^2 \hat{\Sigma}[P, Q], \tag{7.45}$$

with

$$\hat{\Sigma}[P, Q] = 2\int_0^\infty \rho d\rho\, \hat{\mathcal{T}}_{00}(\rho) = \int_0^\infty \rho d\rho \left[\frac{1}{2\rho^2}(\partial_\rho Q)^2 + (\partial_\rho P)^2 + \frac{P^2Q^2}{\rho^2} + \kappa^2(P^2-1)^2\right]. \tag{7.46}$$

In Eq. (7.46), the form of $\hat{\Sigma}[P, Q]$ depends on parameters in the AH model only through κ. Therefore, $P(\rho)$ and $Q(\rho)$ of the vortex solution, and accordingly the dimensionless EMT Eq. (7.40), depend only on κ. Also, the energy per unit length, i.e., the string tension of the vortex, σ_{AH}, is given by

$$\sigma_{\rm AH} = 2\pi v^2 \hat{\sigma}_{\rm AH}(\kappa), \tag{7.47}$$

where $\hat{\sigma}_{\rm AH}(\kappa)$ is obtained by substituting the vortex solution into $\hat{\Sigma}[P, Q]$. As shown in Appendix E, it is possible to show $\hat{\sigma}_{\rm AH}(1/\sqrt{2}) = 1$ analytically [34].

7.3.3 Physical Units

To compare the EMT distribution around the magnetic vortex obtained in the AH model with the flux tube in Chap. 6, it is desirable to introduce a physical dimension into the former. The only parameter having a mass dimension in the AH model is v. To determine this quantity in physical units, we require that the string tension of the magnetic vortex $\sigma_{\rm AH}$ is equivalent with the string tension of the flux tube in the SU(3) YM theory, $\sigma_{\rm YM}$ [18, 25]. From Eq. (7.47), we then have

$$v = \sqrt{\frac{\sigma_{\rm YM}}{2\pi \hat{\sigma}_{\rm AH}(\kappa)}}. \tag{7.48}$$

The value of $\hat{\sigma}_{\rm AH}(\kappa)$ is obtained numerically from the solution of the magnetic vortex with an infinite length. In Fig. 7.1, we show the behavior of $\hat{\sigma}_{\rm AH}(\kappa)$ as a function of κ. The figure shows that $\hat{\sigma}_{\rm AH} = 1$ at $\kappa = 1/\sqrt{2}$. This property can be shown analytically [34], as discussed in Appendix E.

For the value of $\sigma_{\rm YM}$, we use

$$\sigma_{\rm YM} = 1.132(10)\ {\rm GeV/fm}, \tag{7.49}$$

which is obtained from the large R behavior of the $Q\bar{Q}$ potential at $\beta = 6.600$ in Chap. 6.

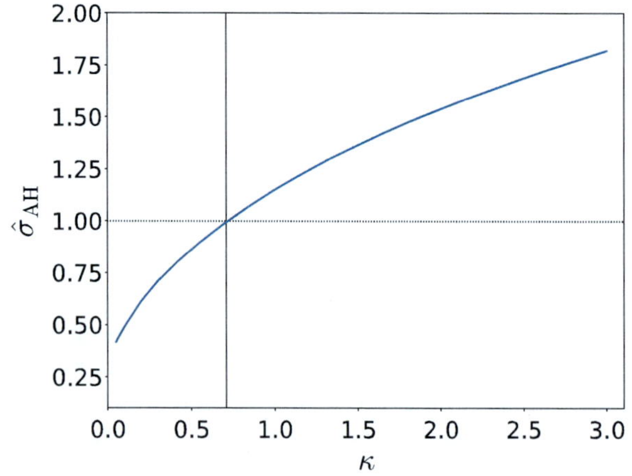

Fig. 7.1 Behavior of $\hat{\sigma}_{\rm AH}(\kappa)$. The vertical line shows the Bogomol'nyi bound $\kappa = 1/\sqrt{2}$

Because the right-hand side of Eq. (7.48) depends only on κ, even after fixing the value of v in physical units, there exists an arbitrariness to vary g and λ with fixed $\kappa = \sqrt{\lambda}/g$. This means that ξ_χ and ξ_A are not determined by fixing v and κ. This arbitrariness is canceled out in Eq. (7.40) for the infinitely long case, but has to be taken into account explicitly when R is finite.

7.3.4 Details of Numerical Analysis

The classical solution of the magnetic vortex is obtained numerically by minimizing the total energy with the boundary condition Eqs. (7.37) and (7.38). For this procedure with finite length R, we discretize the half plane of r and z and iteratively update the fields $\phi(r, z)$ and $\tilde{a}(r, z)$ at even and odd sites via the over-relaxation method. We take the mesh size Δ of the lattice as

$$\Delta = p \min(\xi_A, \xi_\chi), \tag{7.50}$$

where p is chosen in the range $0.02 < p < 0.1$ so that the ratio of R and 2Δ is given by an integer. We have checked that the mesh size with $p = 0.1$ is small enough to suppress the discretization error in the range $0.1 < \kappa < 1.0$ by changing the mesh size. The spatial lengths along the r and z directions are chosen so that the boundaries are at least 3.4 times longer than $\max(\xi_A, \xi_\chi)$. We have checked that the finite size effects are well suppressed with this setting. The iteration is terminated when the total energy becomes unchanged in each step. The criterion is set to be $\delta E = (E_{n+1} - E_n)/E_n < 1.0 \times 10^{-6}$ for the nth iteration. Once we obtain the solution for ϕ and $\tilde{a}$, EMT is obtained by substituting them into Eqs. (7.17)–(7.20).

For the infinitely long case, we proceed similar procedures in the one-dimensional space of ρ for the dimensionless functions $P(\rho)$ and $Q(\rho)$.

We note that the number of lattice points increases when the difference between ξ_A and ξ_χ is large in order to satisfy the above conditions for the mesh size Δ and the size of the lattice. This means that the numerical cost increases at small and large κ. Because of this difficulty, we limit our numerical analysis in the range $0.1 \leq \kappa \leq 1.0$ and $0.05 \leq \kappa \leq 3.0$ for the finite-length and infinitely long vortices, respectively.

7.4 Numerical Results

In this section, we discuss the numerical results on the EMT distribution around the magnetic vortex in the AH model.

7.4.1 Infinitely Long Vortex

In this subsection, we first focus on the infinitely long vortex. In this case, it is convenient to employ the dimensionless variables, ρ and $\hat{\mathcal{T}}_{\gamma\gamma}(\rho)$, introduced in Eqs. (7.39) and (7.40), which depend only on κ. In the left panels of Fig. 7.2, we show the ρ dependences of $\hat{\mathcal{T}}_{00}(\rho)$, $\hat{\mathcal{T}}_{zz}(\rho)$, $\hat{\mathcal{T}}_{rr}(\rho)$, and $\hat{\mathcal{T}}_{\theta\theta}(\rho)$ for three values of κ; the top and lower panels show the results for the type-I ($\kappa = 0.1$) and type-II ($\kappa = 2.0$)

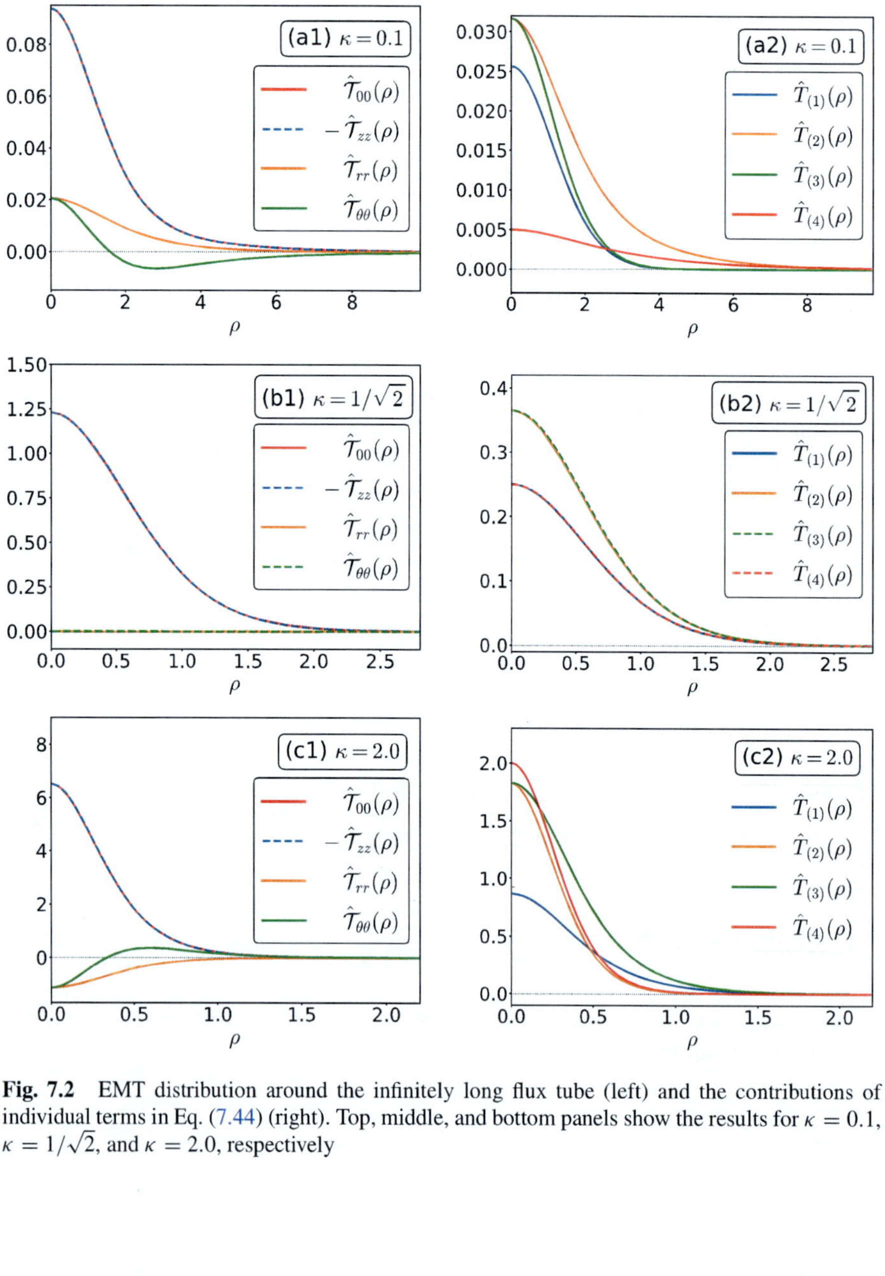

Fig. 7.2 EMT distribution around the infinitely long flux tube (left) and the contributions of individual terms in Eq. (7.44) (right). Top, middle, and bottom panels show the results for $\kappa = 0.1$, $\kappa = 1/\sqrt{2}$, and $\kappa = 2.0$, respectively

cases, respectively, while the middle panel corresponds to the Bogomol'nyi bound at $\kappa = 1/\sqrt{2}$. One finds that the sign of $\mathcal{T}_{rr}(r)$ is positive (negative) at $\kappa = 0.1$ ($\kappa = 2.0$), while the middle panel shows that $\mathcal{T}_{rr}(r) = \mathcal{T}_{\theta\theta}(r) = 0$ at $\kappa = 1/\sqrt{2}$. The latter property is obtained analytically as discussed in Refs. [34, 35] and summarized in Appendix E. For $\kappa \neq 1/\sqrt{2}$, $\mathcal{T}_{\theta\theta}(r)$ behaves differently from $\mathcal{T}_{rr}(r)$ and changes the sign at nonzero r. This result is consistent with the discussion based on the momentum conservation.

Now, let us compare the result in Fig. 7.2 with the EMT distribution around the flux tube in the SU(3) YM theory shown in Fig. 6.11. As in Fig. 6.11, in the SU(3) YM theory $\mathcal{T}_{rr}(r)$ and $\mathcal{T}_{\theta\theta}(r)$ are degenerated within statistics even at the largest $Q\bar{Q}$ distance, $R = 0.92$ fm. The lattice result also suggests that $\mathcal{T}_{\theta\theta}(r)$ is always positive. These results clearly contradict Fig. 7.2. This result thus shows that the structure of the flux tube in the SU(3) YM theory with $R \leq 0.92$ fm cannot be understood by comparison with a magnetic vortex with an infinite length in the AH model.

In spite of this conclusion, it is instructive to take a closer look at the EMT distribution in the AH model in Fig. 7.2. As in Eqs. (7.41)–(7.43), the EMT around the infinitely long vortex consists of four terms in Eq. (7.44). In the right panels of Fig. 7.2, the behavior of these terms is shown separately for each κ. At $\rho = 0$, one has $\hat{T}_{(2)}(\rho) = \hat{T}_{(3)}(\rho)$, which leads to $\mathcal{T}_{rr}(0) = \mathcal{T}_{\theta\theta}(0)$. The sign of $\mathcal{T}_{rr}(0)$ thus is determined by the interplay between $\hat{T}_{(1)}(\rho)$ and $\hat{T}_{(4)}(\rho)$. As discussed in Sect. 7.2.2, $\hat{T}_{(1)}(\rho)$ is the Maxwell stress having a positive contribution to $\mathcal{T}_{rr}(r)$ as in Eq. (7.42), while the contribution of $\hat{T}_{(4)}(\rho)$ is negative in this channel. A positive $\mathcal{T}_{rr}(0)$ at $\kappa < 1/\sqrt{2}$ thus means $\hat{T}_{(1)}(0) > \hat{T}_{(4)}(0)$; i.e., the contribution of the gauge field plays a dominant role at the core of the vortex in the type-I region. On the other hand, for $\kappa > 1/\sqrt{2}$ the effect of the Higgs potential dominates over the gauge field.

In this way, the sign of $\mathcal{T}_{rr}(0)$ can be used to distinguish between type-I and type-II provided that the length of the vortex is infinite. In Fig. 7.3, we show the ratio $\mathcal{T}_{rr}(0)/\mathcal{T}_{00}(0)$ as a function of κ. The figure shows that $\mathcal{T}_{rr}(0)/\mathcal{T}_{00}(0)$ is a monotonic function of κ changing the sign at $\kappa = 1/\sqrt{2}$.

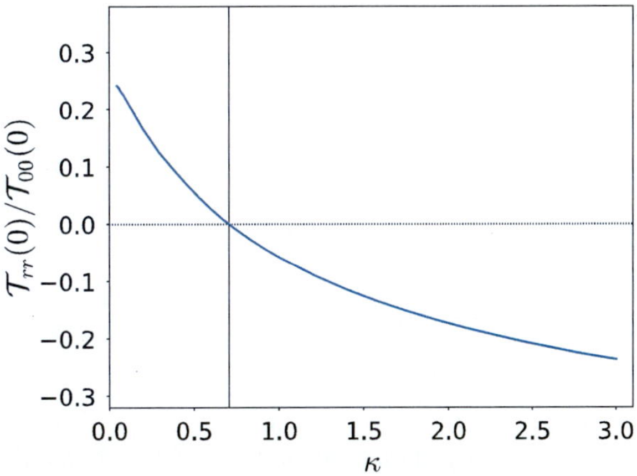

Fig. 7.3 κ-dependence of the ratio $\mathcal{T}_{rr}(0)/\mathcal{T}_{00}(0)$ for the infinitely long vortex. The value $\kappa = 1/\sqrt{2}$ is indicated by the vertical line

7.4.2 Finite-Length Flux Tube

Next, we investigate a magnetic vortex with finite length R. In the following, numerical results are shown in physical units by fixing the value of v through Eq. (7.48) in order to make the comparison with the results in Chap. 6 easy. After fixing the value of v in physical units, the vortex solution with length R depends on two parameters in the AH model. In the following, we use κ and g as the parameters.

We first fix the length of the vortex to be $R = 0.92$ fm, the largest length of the flux tube in Chap. 6, and study the κ and g dependence of the EMT distribution on the mid-plane. Shown in Fig. 7.4 are the EMT distributions on the mid-plane for various combinations of κ and g. The values of κ and g increase along the right and lower

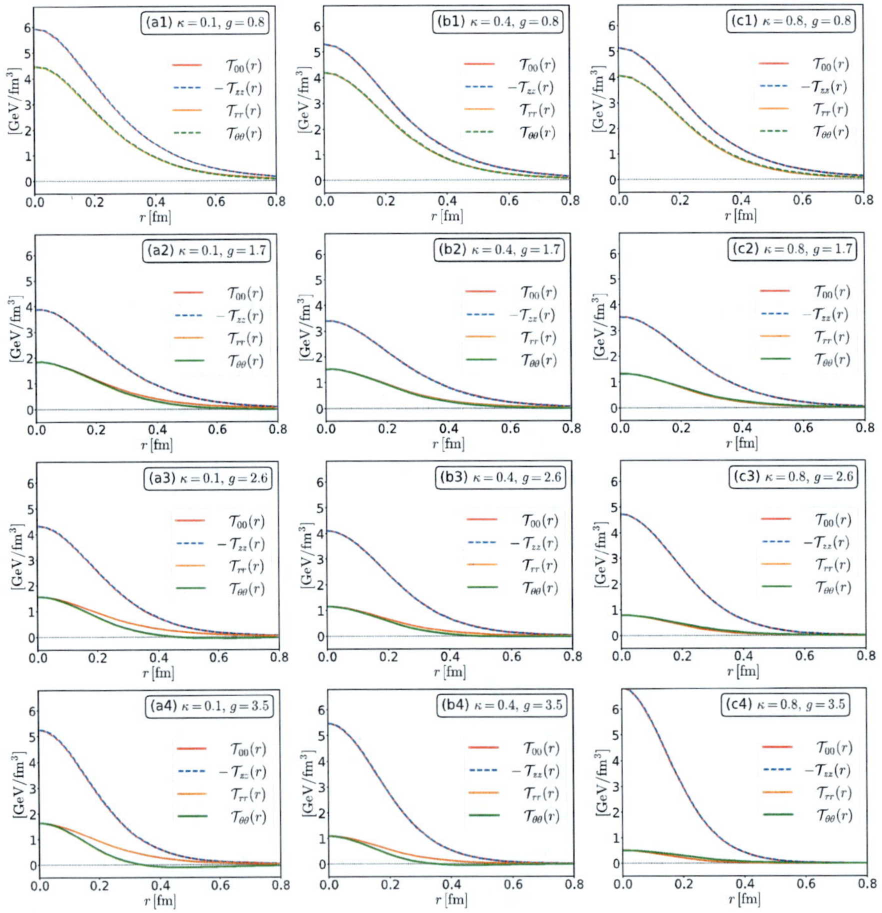

Fig. 7.4 EMT distribution on the mid-plane of the magnetic vortex with $R = 0.92$ fm for various combinations of κ and g. The left, middle, and right panels show the results for $\kappa = 0.1$, 0.4, and 0.8, respectively, while the value of g is $g = 0.8,\ 1,\ 7,\ 2.6$, and 3.5 from the top to the bottom

Table 7.1 Parameters κ and g of the numerical analysis in Fig. 7.4, and the corresponding values of the correlation lengths ξ_χ and ξ_A

g \ κ		0.1	0.4	0.8
0.8	ξ_χ [fm]	4.57	1.45	0.84
	ξ_A [fm]	0.65	0.82	0.94
1.7	ξ_χ [fm]	2.15	0.68	0.39
	ξ_A [fm]	0.30	0.39	0.45
2.6	ξ_χ [fm]	1.41	0.45	0.26
	ξ_A [fm]	0.20	0.25	0.29
3.5	ξ_χ [fm]	1.04	0.33	0.19
	ξ_A [fm]	0.15	0.19	0.22

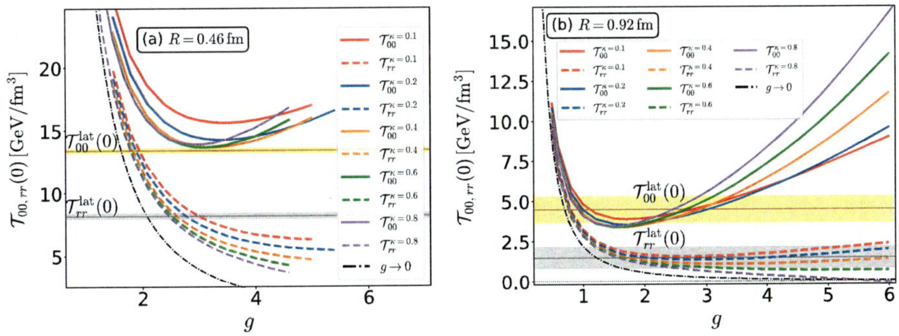

Fig. 7.5 Dependences of $\mathcal{T}_{00}(0)$ and $\mathcal{T}_{rr}(0)$ on g for several values of κ at $R = 0.46$ fm (left) and $R = 0.92$ fm (right). The dash-dotted line shows the contribution from the gauge field $\mathbf{A}^D(r, z)$, Eq. (7.28)

directions, respectively. The correlation lengths, ξ_χ and ξ_A, corresponding to each panel, are shown in Table 7.1. With fixed κ, ξ_χ and ξ_A are monotonically decreasing as g becomes larger as in Eq. (7.11). The effect of boundaries thus becomes smaller as g becomes larger. By increasing κ with fixed g, on the other hand, ξ_χ increases but ξ_A decreases. This behavior comes from the κ dependence of v in Eq. (7.48).

From Fig. 7.4, one finds that the difference between $\mathcal{T}_{rr}(r)$ and $\mathcal{T}_{\theta\theta}(r)$ tends to decrease as g becomes smaller. In particular, one sees that these channels are almost degenerated in the upper two rows, while these channels have a clear separation from $\mathcal{T}_{00}(r)$ and $\mathcal{T}_{zz}(r)$. This result suggests that the degeneracy between $\mathcal{T}_{rr}(r)$ and $\mathcal{T}_{\theta\theta}(r)$ and their separation from $\mathcal{T}_{00}(r)$ and $\mathcal{T}_{zz}(r)$ observed in Chap. 6 can be described by the effects of the boundaries.

Next, to make a more quantitative comparison between the vortex in the AH model and the lattice results, we focus on the absolute values of $\mathcal{T}_{00}(r)$ and $\mathcal{T}_{rr}(r)$ at $r = 0$. In Fig. 7.5, we show the values of $\mathcal{T}_{00}(0)$ and $\mathcal{T}_{rr}(0)$ as functions of g for several values of κ at $R = 0.46$ and 0.92 fm. In Fig. 7.5, we also plot Eq. (7.28) with

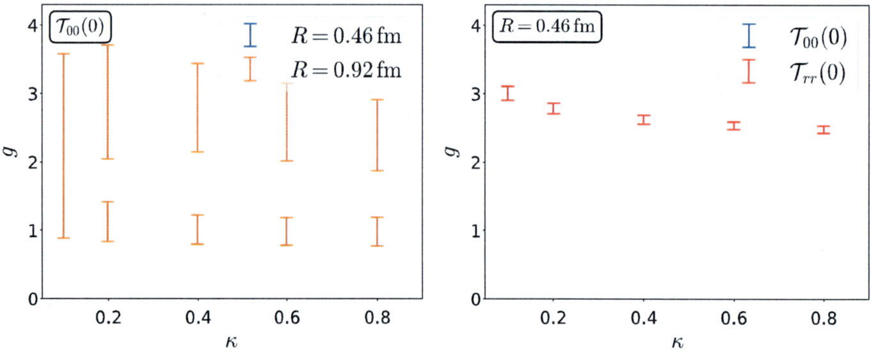

Fig. 7.6 (Left) Range of the parameter g which gives the value of $\mathcal{T}_{00}(0)$ consistent with the lattice result in Chap. 6. (Right) Range of the parameter g which gives the values of $\mathcal{T}_{00}(0)$ and $\mathcal{T}_{rr}(0)$ consistent with the lattice result in Chap. 6 at $R = 0.46$ fm

Table 7.2 Values of $\mathcal{T}_{00}(0)$ and $\mathcal{T}_{rr}(0)$ in Chap. 6 for flux tubes with lengths $R = 0.46$ and 0.92 fm

R [fm]	$\mathcal{T}_{00}(0)$ [GeV/fm^3]	$\mathcal{T}_{rr}(0)$ [GeV/fm^3]
0.46	13.4 (2)	8.2 (2)
0.92	4.5 (9)	1.5 (7)

$n = 1$ by the dash-dotted lines. When the condition $R \ll \xi_A, \xi_\chi$ is satisfied, EMT is expected to be dominated by the contribution from the gauge field $\mathbf{A}^D(r, z)$ as discussed in Sect. 7.3.1. In this case, which is realized in the small g limit, $\mathcal{T}_{00}(0)$ and $\mathcal{T}_{rr}(0)$ should approach as in Eq. (7.28). The figure shows that $\mathcal{T}_{00}(0)$ and $\mathcal{T}_{rr}(0)$ have steep rises corresponding to Eq. (7.28) at small g. From the figure, one also finds that $\mathcal{T}_{00}(0)$ and $\mathcal{T}_{rr}(0)$ go toward degeneracy in this g range.

On the other hand, in $R \gg \xi_A, \xi_\chi$, which indicates the large g limit, the asymptotic behavior of each component of EMT is constrained by Eq. (7.40) as follows:

$$\mathcal{T}_{\gamma\gamma'}(0) = 2g^2 v^4 \hat{\mathcal{T}}_{\gamma\gamma'}(0) = g^2 \frac{\sigma_{\text{YM}}^2}{2\pi^2 \hat{\sigma}_{\text{AH}}^2} \hat{\mathcal{T}}_{\gamma\gamma'}(0), \tag{7.51}$$

where $\hat{\mathcal{T}}_{\gamma\gamma'}(0)$ is obtained from the analyses of the infinitely long vortices. Therefore, the asymptotic value of each components depends on κ. In particular, in the case of type-II, $\mathcal{T}_{rr}(0)$ approaches a negative value.

The horizontal lines in Fig. 7.5 show the values of $\mathcal{T}_{00}(0)$ and $\mathcal{T}_{rr}(0)$ in Chap. 6 with the error bars indicated by the shaded region; the numerical values of $\mathcal{T}_{00}(0)$ and $\mathcal{T}_{rr}(0)$ are given in Table 7.2. We note that the value of $\mathcal{T}_{00}(0)$ used here is obtained from the average $(\mathcal{T}_{00}(0) + \mathcal{T}_{zz}(0))/2$ in Chap. 6. The value of g with fixed κ can be constrained by requiring that $\mathcal{T}_{00}(0)$ of the magnetic vortex reproduces these lattice results. In the left panel of Fig. 7.6, we show the range of g determined in this way for $R = 0.46$ and 0.92 fm by the bands. The upper and lower bounds of the bands in the panel are the values of g at which $\mathcal{T}_{00}(0)$ of the vortex is the upper and lower bounds of the error bar of the lattice result. We note that there are two ranges of g for $R = 0.92$ fm and $\kappa \geq 0.2$ because of the non-monotonic behavior of $\mathcal{T}_{00}(0)$ as a function of g as shown in Fig. 7.5. In the result with $R = 0.46$ fm, the value of $\mathcal{T}_{00}(0)$ never agrees with the lattice result in Chap. 6 within the statistical error. In the left panel of Fig. 7.6, therefore, the data corresponding to $\mathcal{T}_{00}(0)$ at $R = 0.46$ fm are not shown. This result shows that the vortex solution in the AH model does not have a parameter set which reproduces the EMT distribution of the flux tube in the SU(3) YM theory for $0.1 < \kappa < 0.8$. The same conclusion is also obtained from the right panel of Fig. 7.6 that shows the range of g constrained by requiring that $\mathcal{T}_{00}(0)$ and $\mathcal{T}_{rr}(0)$ are consistent with the lattice results at $R = 0.46$ fm.

References

1. Suzuki H (2013) Energy-momentum tensor from the Yang-Mills gradient flow. PTEP 2013(8):083B03 [Erratum: PTEP 20157:079201 (2015)]. arXiv:1304.0533 [hep-lat]
2. Asakawa M et al (2014) [FlowQCD], Thermodynamics of SU(3) gauge theory from gradient flow on the lattice. Phys Rev D 90(1):011501 [erratum: Phys Rev D 92(5):059902 (2015)]. arXiv:1312.7492 [hep-lat]
3. Makino H, Suzuki H (2014) Lattice energy–momentum tensor from the Yang–Mills gradient flow—inclusion of fermion fields. PTEP 2014:063B02 [erratum: PTEP 2015:079202 (2015)]. arXiv:1403.4772 [hep-lat]
4. Kitazawa M, Iritani T, Asakawa M, Hatsuda T, Suzuki H (2016) Equation of state for SU(3) Gauge theory via the energy-momentum tensor under gradient flow. Phys Rev D 94(11):114512. arXiv:1610.07810 [hep-lat]
5. Taniguchi Y, Ejiri S, Iwami R, Kanaya K, Kitazawa M, Suzuki H, Umeda T, Wakabayashi N (2017) Exploring N_f = 2+1 QCD thermodynamics from the gradient flow. Phys Rev D 96(1):014509 [erratum: Phys Rev D 99(5):059904 (2019)]. arXiv:1609.01417 [hep-lat]
6. Taniguchi Y et al (2020) [WHOT-QCD], N_f = 2+1 QCD thermodynamics with gradient flow using two-loop matching coefficients. Phys Rev D 102(1):014510. arXiv:2005.00251 [hep-lat]
7. Narayanan R, Neuberger H (2006) Infinite N phase transitions in continuum Wilson loop operators. JHEP 0603:064. [hep-th/0601210]
8. Lüscher M (2010) Properties and uses of the Wilson flow in lattice QCD. JHEP 1008:071. arXiv:1006.4518 [hep-lat]
9. Lüscher M, Weisz P (2011) Perturbative analysis of the gradient flow in non-abelian gauge theories. JHEP 1102:051. arXiv:1101.0963 [hep-th]
10. Nambu Y (1974) Strings, monopoles and Gauge fields. Phys Rev D 10:4262
11. 't Hooft G Gauge fields with unified weak, electromagnetic, and strong interactions. PRINT-75-0836 (UTRECHT)
12. Mandelstam S (1976) Vortices and quark confinement in nonabelian Gauge theories. Phys Rept 23:245
13. Nielsen HB, Olesen P (1973) Vortex line models for dual strings. Nucl Phys B 61:45
14. 't Hooft G (1981) Topology of the Gauge condition and new confinement phases in nonabelian Gauge theories. Nucl Phys B 190:455
15. Suzuki T (1988) A Ginzburg-Landau type theory of quark confinement. Prog Theor Phys 80:929
16. Ball JS, Caticha A (1988) Superconductivity: a testing ground for models of confinement. Phys Rev D 37:524

17. Maedan S, Matsubara Y, Suzuki T (1990) Abelian confinement mechanism and the QCD vacuum. Prog Theor Phys 84:130
18. Kodama H, Matsubara Y, Ohno S, Suzuki T (1997) Inter - meson potentials in dual Ginzburg-Landau theory. Prog Theor Phys 98:1345
19. Ezawa ZF, Iwazaki A (1982) Abelian dominance and quark confinement in Yang-Mills theories. Phys Rev D 25:2681
20. Ezawa ZF, Iwazaki A (1982) Abelian dominance and quark confinement in Yang-Mills theories. 2. Oblique confinement and η' mass. Phys Rev D 26:631
21. Suzuki T, Yotsuyanagi I (1990) A possible evidence for Abelian dominance in quark confinement. Phys Rev D 42:4257–4260
22. Kronfeld AS, Schierholz G, Wiese UJ (1987) Topology and dynamics of the confinement mechanism. Nucl Phys B 293:461
23. Kronfeld AS, Laursen ML, Schierholz G, Wiese UJ (1987) Monopole condensation and color confinement. Phys Lett B 198:516
24. Maedan S, Suzuki T (1989) An infrared effective theory of quark confinement based on monopole condensation. Prog Theor Phys 81:229
25. Suganuma H, Sasaki S, Toki H (1995) Color confinement, quark pair creation and dynamical chiral symmetry breaking in the dual Ginzburg-Landau theory. Nucl Phys B 435:207–240. arXiv:hep-ph/9312350 [hep-ph]
26. Sasaki S, Suganuma H, Toki H (1995) Dual Ginzburg-Landau theory with QCD monopoles for dynamical chiral symmetry breaking. Prog Theor Phys 94:373–384
27. Ichie H, Suganuma H, Toki H (1995) QCD phase transition at finite temperature in the dual Ginzburg-Landau theory. Phys Rev D 52:2944–2950. arXiv:hep-ph/9502278 [hep-ph]
28. Monden H, Suganuma H, Ichie H, Toki H (1998) Surface tension in the mixed phase of QCD in the dual Ginzburg-Landau theory. Phys Rev C 57:2564–2575. arXiv:hep-ph/9701271 [hep-ph]
29. Koma Y, Suganuma H, Toki H (1999) Flux tube ring and glueball properties in the dual Ginzburg-Landau theory. Phys Rev D 60:074024. arXiv:hep-ph/9902441 [hep-ph]
30. Koma Y, Koma M, Ilgenfritz EM, Suzuki T, Polikarpov MI (2003) Duality of gauge field singularities and the structure of the flux tube in Abelian projected SU(2) gauge theory and the dual Abelian Higgs model. Phys Rev D 68:094018. arXiv:hep-lat/0302006 [hep-lat]
31. Oxman LE (2017) $SU(N) \to Z(N)$ dual superconductor models: the magnetic loop ensemble point of view. EPJ Web Conf 137:03015. arXiv:1704.01989 [hep-th]
32. Oxman LE (2018) 4D ensembles of percolating center vortices and monopole defects: the emergence of flux tubes with N -ality and gluon confinement. Phys Rev D 98(3):036018. arXiv:1805.06354 [hep-th]
33. Ginzburg VL, Landau LD (1950) On the theory of superconductivity. Zh Eksp Teor Fiz 20:1064
34. Bogomol'nyi EB (1976) The stability of classical solutions. Sov J Nucl Phys 24:449 PRINT-76-0543 (LANDAU-INST)
35. de Vega HJ, Schaposnik FA (1976) A classical vortex solution of the Abelian Higgs model. Phys Rev D 14:1100–1106
36. Fetter AL, Walecka JD (1971) Quantum theory of many-particle systems. McGraw-Hill
37. Dirac PAM (1931) Quantised singularities in the electromagnetic field. Proc Roy Soc Lond A 133(821):60–72

Chapter 8
Summary and Outlook

In this dissertation, we have investigated the distributions of the energy momentum tensor (EMT) around static charges both in the lattice simulations based on the SU(3) Yang–Mills (YM) gauge theory and in the Abelian-Higgs (AH) model. In Chaps. 5 and 6, we have investigated the EMT distributions in two kinds of systems of the SU(3) YM gauge theory on the lattice: One is the static single quark (Q) system in the deconfined phase and the other is the static quark–anti-quark ($Q\bar{Q}$) system in vacuum. In these studies, the YM gradient flow makes it possible to define the EMT operator and apply to the numerical simulation on the lattice. In Chap. 7, motivated by the lattice study in Chap. 6, we have studied the EMT distribution around $Q\bar{Q}$ in terms of the conservation law of EMT as well as the AH model.

Throughout our studies, EMT plays crucial roles as a direct and fundamental physical observable to explore such complex quantum systems as the YM gauge theory and provide us deep insights into the systems. We have employed static test charges as probes for the exploration of such systems. We have obtained the response of the systems against those static charges through the gauge invariant observable, EMT, and then have revealed the local properties of the vacuum as well as the hot medium of the YM gauge theory.

In Chap. 5, we have investigated the spatial distribution of EMT around a static Q at finite temperature above the critical temperature T_c of the SU(3) YM gauge theory on the lattice. We found that there are no significant differences between the magnitudes of the resultant EMT distributions in the radial and angular directions for all the temperatures. In addition, the sign of each EMT component is the same as that in the Maxwell theory. In this sense, the results are in accordance with the prediction in the Maxwell theory . However, we found the substantial difference of the temporal component from the spatial ones, which becomes more remarkable near T_c. This result suggests the genuine non-Abelian effect at finite temperature.

R. Yanagihara, *Distribution of Energy Momentum Tensor around Static Charges in Lattice Simulations and an Effective Model*, Springer Theses,
https://doi.org/10.1007/978-981-16-6234-8_8

We then have investigated the temperature dependence of the distribution of EMT. At short distance, $r \lesssim 0.2$ fm, all the EMT distributions show small temperature dependence. Based on the results in the leading-order (LO) perturbation in the electrostatic QCD, we have extracted the strong coupling constant $\alpha_s(r)$ from the ratio of each EMT component to the trace of EMT. We found that $\alpha_s(r = 0.1\,\mathrm{fm}) \simeq 0.22 - 0.32$. Although the results have channel dependence, indicating the importance of the higher-order contributions in perturbation, they are almost consistent with that obtained from the similar analysis for the $Q\bar{Q}$ free energy at finite temperature.

On the other hand, at large distance, $r \gtrsim 0.2$ fm, we found the sizable temperature dependence of the EMT distribution. Although the thermal Debye screening should be studied in this region, the large statistical errors prevent us from quantitatively extracting the information on it.

In Chap. 6, we have studied the EMT distribution around a static $Q\bar{Q}$ pair in vacuum of the SU(3) YM gauge theory on the lattice. Before taking the double extrapolation, the distribution of EMT decomposed into local principle axes qualitatively reveals the spatial structure of the flux tube in a gauge-invariant manner.

On the mid-plane between $Q\bar{Q}$, we have analyzed the EMT distribution quantitatively by taking the double extrapolation. We found the approximate degeneracies $\langle \mathcal{T}^{\mathrm{R}}_{44}(r)\rangle_{Q\bar{Q}} \simeq \langle \mathcal{T}^{\mathrm{R}}_{zz}(r)\rangle_{Q\bar{Q}} < 0$ as well as $\langle \mathcal{T}^{\mathrm{R}}_{rr}(r)\rangle_{Q\bar{Q}} \simeq \langle \mathcal{T}^{\mathrm{R}}_{\theta\theta}(r)\rangle_{Q\bar{Q}} > 0$ for a wide range of the radial direction r and the $Q\bar{Q}$ distance R while the magnitude of the former is larger than the latter. The nonzero trace of EMT is seen, which indicates the partial restoration of the scale symmetry inside the flux tube. In addition, we have estimated that the typical transverse size of the flux tube is $\sim$0.2 fm and it becomes larger with the increase of R. The quantitative determination of the radius is prevented from the lack of enough statistics.

We have also confirmed the agreement between the force acting on the charge obtained from the derivative of the $Q\bar{Q}$ potential with respect to R and the force evaluated by the surface integral of the stress tensor surrounding the charge. This consistency in the results supports the validity of our analysis.

In Chap. 7, in order to further investigate the local structure of the flux tube in Chap. 6, we have studied the EMT distribution around static $Q\bar{Q}$ from the viewpoint of the conservation law of EMT and by employing the AH model. From the momentum conservation, we found that the lattice results on the mid-plane obtained in Chap. 6 are qualitatively inconsistent with the assumption of the translational invariance along the z direction even at $R = 0.92$ fm. This indicates the importance of boundaries corresponding to the $Q\bar{Q}$ charges when we perform any model analyses on the flux tube in the SU(3) YM theory, whose length is less than 0.92 fm.

We have then employed the AH model and calculated the EMT distribution around the magnetic vortex on the mid-plane between a monopole and an anti-monopole. The resultant distribution with a finite-length vortex shows the degeneracy between the radial and angular components, and their separation from the temporal component. This is qualitatively consistent with the result in the lattice study. However, we found that a wide range of the parameters in the AH model cannot quantitatively reproduce the EMT distribution in Chap. 6, which is revealed from the comparison of the magnitude of both temporal and radial components of EMT.

In our studies, we have revealed the non-perturbative aspects of the SU(3) YM gauge theory with the use of EMT. We have a lot of important issues beyond these studies presented in this dissertation. For the study in Chap. 5, the higher-order thermal QCD calculation of the EMT distribution around a single Q is called for to understand the lattice results [1]. In order to precisely extract the Debye mass from the EMT distribution at large distance, we need to increase the statistics of the lattice data. In Chap. 6, it is an important future work to calculate the width of the flux tube with larger amounts of statistics, which is predicted to fatten logarithmically as a function of R [2, 3]. For the comparison with the lattice results, such perturbative calculations of the EMT distribution around static color charges as those with the careful treatment of the renormalons are interesting subjects, which have succeeded in the calculation of the QCD potential [4–6].

The extension to the full QCD calculation is an interesting future problem for our present studies given in Chaps. 5 and 6. There are many open questions on this point: the single Q system below/above T_c, the QQ and $Q\bar{Q}$ systems at finite temperature [7], the QQQ system [8], and so on. In particular, the single Q system in QCD below T_c corresponds to a heavy-light meson [9]. There is possibility for analyses on the QQQ system to provide new insights into the flux tube formation in baryons. It further leads to information on the hadron structure, such as the gravitational form factor [10–14]. These studies can be performed based on essentially the same formalism as utilized in Chaps. 5 and 6.

The improvement of the analysis with the AH model in Chap. 7 is an important issue although not all the parameter regions have been excluded yet in the AH model employed above. One possibility is to consider the quantum effect in the AH model. Such a study contributes to estimate how the quantum fluctuation affects the width of the flux tube in addition to the thickness of the flux tube itself [15].

References

1. Private communication with Berwein M
2. Lüscher M, Munster G, Weisz P (1981) How thick are chromoelectric flux tubes? Nucl Phys B 180:1–12
3. Gliozzi F, Pepe M, Wiese UJ (2010) The width of the color flux tube at 2-loop order. JHEP 11:053. arXiv:1006.2252 [hep-lat]
4. Sumino Y Understanding interquark force and quark masses in perturbative QCD. arXiv:1411.7853 [hep-ph]
5. Sumino Y (2002) A Connection between the perturbative QCD potential and phenomenological potentials. Phys Rev D 65:054003. arXiv:hep-ph/0104259 [hep-ph]
6. Recksiegel S, Sumino Y (2002) Perturbative QCD potential, renormalon cancellation and phenomenological potentials. Phys Rev D 65:054018. arXiv:hep-ph/0109122 [hep-ph]
7. Maezawa Y, Umeda T, Aoki S, Ejiri S, Hatsuda T, Kanaya K, Ohno H (2012) Application of fixed scale approach to static quark free energies in quenched and 2+1 flavor lattice QCD with improved Wilson quark action. Prog Theor Phys 128:955–970. arXiv:1112.2756 [hep-lat]

8. Takahashi TT, Suganuma H, Nemoto Y, Matsufuru H (2002) Detailed analysis of the three quark potential in SU(3) lattice QCD. Phys Rev D 65:114509. arXiv:hep-lat/0204011 [hep-lat]
9. Müller L, Philipsen O, Reisinger C, Wagner M (2019) Hybrid static potential flux tubes from SU(2) and SU(3) lattice gauge theory. Phys Rev D 100(5):054503. arXiv:1907.01482 [hep-lat]
10. Kumano S, Song QT, Teryaev OV (2018) Hadron tomography by generalized distribution amplitudes in pion-pair production process $\gamma^*\gamma \to \pi^0\pi^0$ and gravitational form factors for pion. Phys Rev D 97(1):014020. arXiv:1711.08088 [hep-ph]
11. Polyakov MV, Schweitzer P (2018) Forces inside hadrons: pressure, surface tension, mechanical radius, and all that. Int J Mod Phys A 33(26):1830025. arXiv:1805.06596 [hep-ph]
12. Lorcé C, Moutarde H, Trawiński AP (2019) Revisiting the mechanical properties of the nucleon. Eur Phys J C 79(1):89. arXiv:1810.09837 [hep-ph]
13. Burkert VD, Elouadrhiri L, Girod FX (2018) The pressure distribution inside the proton. Nature 557(7705):396–399
14. Shanahan PE, Detmold W (2019) Pressure distribution and shear forces inside the proton. Phys Rev Lett 122(7):072003. arXiv:1810.07589 [nucl-th]
15. Ito H, Kitazawa M in progress

Appendix A
Improved Estimator

In actual numerical simulations, we seek a method to measure physical quantities with small uncertainty while the expectation value is kept. When we consider the expectation value of the observable A, that is, $\langle A \rangle$, we have an option to select another observable $A_{\rm I}$, such that

$$\langle A_{\rm I} \rangle = \langle A \rangle, \quad \langle A_{\rm I}^2 \rangle < \langle A^2 \rangle. \tag{A.1}$$

It is true that the mean values of A and $A_{\rm I}$ are to be the same if we have infinite statistics, but the uncertainty of $A_{\rm I}$ is smaller than that of A within finite statistics. We refer to such an operator $A_{\rm I}$ as an improved estimator.

In this appendix, we introduce the so-called multihit algorithm in numerical simulations based on the Monte Carlo method [1]. In particular, we consider the improvement of temporal link variables in pure gauge theory. The essential idea is to replace a temporal link variable by the averaged one as follows:

$$U_4(n) \to V_4(n) = \frac{\displaystyle\int dU\, U_4 e^{-\beta S_{n,4}[U]}}{\displaystyle\int dU\, e^{-\beta S_{n,4}[U]}}, \tag{A.2}$$

where

$$S_{n,\mu}[U] = -\frac{1}{2}\mathrm{Tr}[U F_\mu^\dagger(n)], \tag{A.3}$$

$$F_\mu(n) = \sum_{\nu\neq\mu} U_\nu(n) U_\mu(n+\hat{\nu}) U_\nu^\dagger(n+\hat{\mu}). \tag{A.4}$$

In this replacement, $V_4(n)$ no longer belongs to the original gauge group. The replacement with the mean value plays a role of noise reduction.

R. Yanagihara, *Distribution of Energy Momentum Tensor around Static Charges in Lattice Simulations and an Effective Model*, Springer Theses,
https://doi.org/10.1007/978-981-16-6234-8

When we consider the SU(2) gauge theory, we can analytically calculate the integration in Eq. (A.2) as

$$V_4(n) = \frac{I_2(\beta f_\mu(n))}{f_\mu(n) I_1(\beta f_\mu(n))} F_\mu(n), \tag{A.5}$$

where I_n represents the first-order modified Bessel function and $f_\mu(n)=\sqrt{\det(F_\mu(n))}$. In the case of the SU(3) gauge theory, we also explicitly write down the formula despite the lengthy form. In practice, we numerically evaluate Eq. (A.2) by performing the local heat bath update algorithm. The multihit algorithm is named after the iterative application of the heat bath update for the link variables.

We comment that the idea of the improved estimator is seen in the analyses of the Ising model, where one makes use of the Callen identity.

Reference

1. Parisi G, Petronzio R, Rapuano F (1983) A measurement of the string tension near the continuum limit. Phys Lett 128B:418

Appendix B
Tree-Level Improvement of Lattice Observables

As shown in Fig. 5.2, there exists sizable discretization effect for the EMT distribution on coarse lattices especially for $N_\tau = 10$. In this appendix, we attempt to reduce such discretization effects by using the tree-level lattice propagator. A similar idea has been applied to the analysis of the Polyakov loop correlations in Refs. [1, 2].

In calculating the EMT distribution around a static-quark Eq. (5.8), we need the expectation values of Eqs. (4.139) and (4.140) at nonzero flow time t. These quantities are constructed from $\langle G^a_{\mu\nu}(t,\boldsymbol{x})G^a_{\rho\sigma}(t,\boldsymbol{x})\rangle_Q$, where the temporal coordinate is suppressed for notational simplicity. In the continuum theory at the tree level, this operator is calculated to be

$$\langle G^a_{\mu\nu}(t,\boldsymbol{x})G^a_{\rho\sigma}(t,\boldsymbol{x})\rangle_Q = -g^2\frac{N^2-1}{2}\mathcal{G}_{\mu\nu}(t,\boldsymbol{x})\mathcal{G}_{\rho\sigma}(t,\boldsymbol{x}), \tag{B.1}$$

where g is the gauge coupling and the number of color $N = 3$ is the number of colors, with

$$\mathcal{G}_{\mu\nu}(t,\boldsymbol{x})\delta^{ab} = \int_0^{1/T} d\tau\, \langle A^a_4(\tau,\mathbf{0})G^b_{\mu\nu}(t,x)\rangle. \tag{B.2}$$

By selecting appropriate gauge fixing conditions for the gauge action and the gradient-flow equation, one obtains $\mathcal{G}_{12}(t,\boldsymbol{x}) = \mathcal{G}_{23}(t,\boldsymbol{x}) = \mathcal{G}_{31}(t,\boldsymbol{x}) = 0$ and

$$\mathcal{G}_{i4}(t,\boldsymbol{x}) = \partial_i D(t,\boldsymbol{x}), \tag{B.3}$$

$$D(t,\boldsymbol{x}) = \int \frac{d^3p}{(2\pi)^3} e^{ip\cdot x}\frac{e^{-tp^2}}{p^2} = \frac{1}{4\pi|\boldsymbol{x}|}\mathrm{erf}\left(\frac{|\boldsymbol{x}|}{\sqrt{4t}}\right). \tag{B.4}$$

Next, in lattice gauge theory the propagator corresponding to Eq. (B.4) with the Wilson gauge action for the gauge action and the flow equation reads [3, 4]

R. Yanagihara, *Distribution of Energy Momentum Tensor around Static Charges in Lattice Simulations and an Effective Model*, Springer Theses,
https://doi.org/10.1007/978-981-16-6234-8

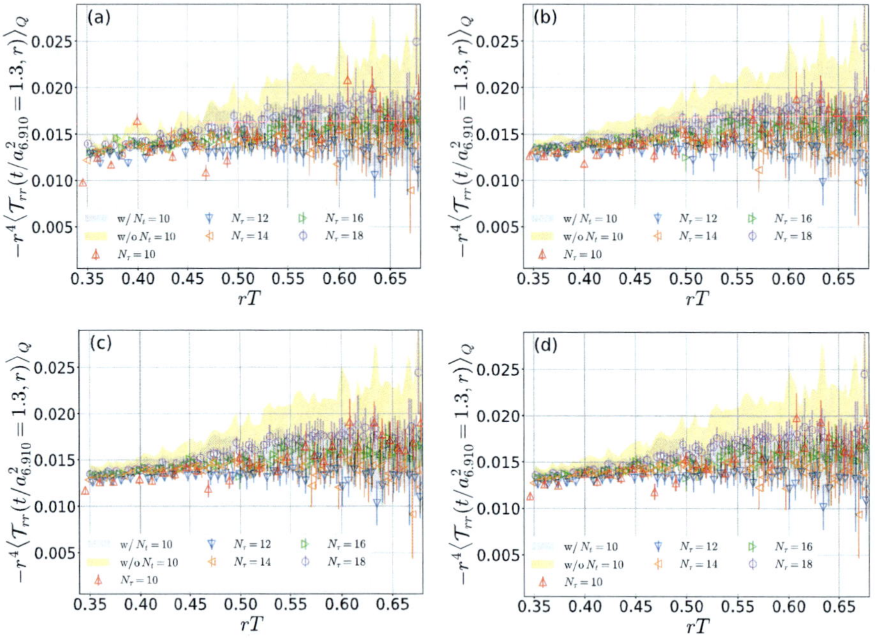

Fig. B.1 Distribution of $-r^4\langle \mathcal{T}_{rr}(t/a_{6.910}^2 = 1.3, r)\rangle_Q$ as functions of rT. (**a**) Same figure as Fig. 5.2 shown for a comparison. (**b**)–(**d**) Tree-level improved results with Eqs. (B.8) and (B.10) for three choices of $U_{\gamma\gamma'}^{\text{tl}}$: (**b**) $U_{\gamma\gamma'}^{\text{tl}} = \langle U_{44}\rangle_Q$, (**c**) $U_{\gamma\gamma'}^{\text{tl}} = \langle U_{rr}\rangle_Q$, (**d**) $U_{\gamma\gamma'}^{\text{tl}} = \langle U_{\theta\theta}\rangle_Q$

$$\mathcal{D}(t, \boldsymbol{x}_n) = \int_{-\pi}^{\pi} \frac{d^3 p}{(2\pi)^3} e^{i\boldsymbol{p}\cdot\boldsymbol{x}_n} \frac{e^{-t\sum_i \hat{p}_i^2}}{\sum_i \hat{p}_i^2}, \tag{B.5}$$

with $\boldsymbol{x}_n = a\boldsymbol{n} = a(n_x, n_y, n_z)$ and $\hat{p}_i = (2/a)\sin(p_i/2a)$. When the clover-leaf operator for the discretized representation of $G_{\mu\nu}^a(t, \boldsymbol{x}_n)$ is employed, the discretized representation of $\mathcal{G}_{i4}(t, \boldsymbol{x}_n)$ is given by [5]

$$\mathcal{G}_{i4}^{\text{lat}}(t, \boldsymbol{x}_n) = \frac{1}{2a}\left(\mathcal{D}(t, \boldsymbol{x}_{n+\hat{i}}) - \mathcal{D}(t, \boldsymbol{x}_{n-\hat{i}})\right). \tag{B.6}$$

Using Eqs. (B.3) and (B.6), the tree-level improvements of Eqs. (4.139) and (4.140) denoted by the superscript 'imp' may be written as

$$\langle U_{\gamma\gamma'}(t, \boldsymbol{x}_n)\rangle_Q^{\text{imp}} = c(t, \boldsymbol{x}_n)\langle U_{\gamma\gamma'}(t, \boldsymbol{x}_n)\rangle_Q, \tag{B.7}$$

$$\langle E(t, \boldsymbol{x}_n)\rangle_Q^{\text{imp}} = c(t, \boldsymbol{x}_n)\langle E(t, \boldsymbol{x}_n)\rangle_Q, \tag{B.8}$$

where the correction factor $c(t, \boldsymbol{x}_n)$ is defined by

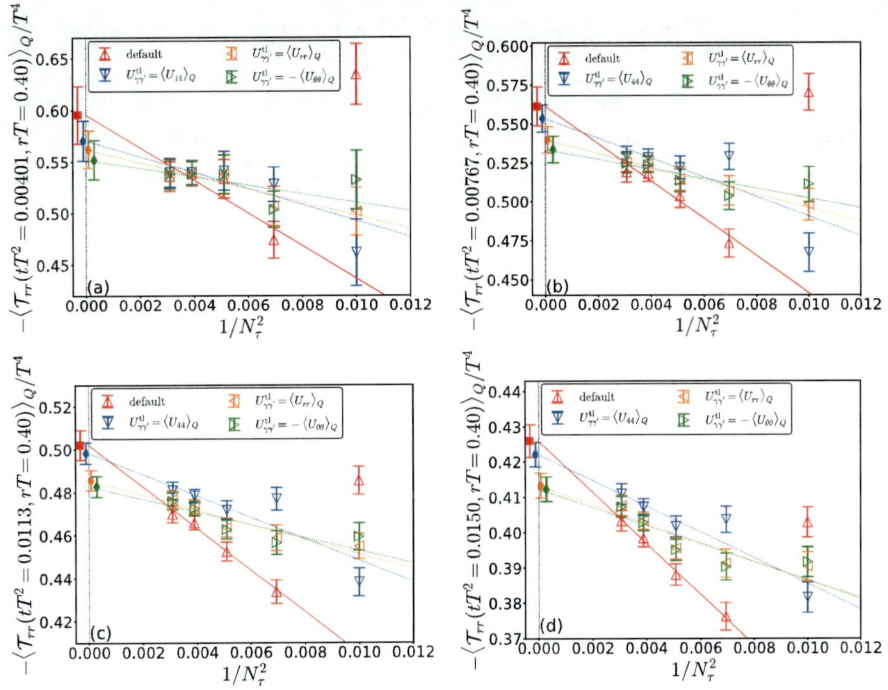

Fig. B.2 Comparison of the different prescriptions in the continuum extrapolation of $-\langle \mathcal{T}_{rr}(t, rT = 0.40)\rangle_Q/T^4$ at $T/T_c = 1.44$: (**a**) $tT^2 = 0.00401$, (**b**) $tT^2 = 0.00767$, (**c**) $tT^2 = 0.0113$, and (**d**) $tT^2 = 0.0150$

$$c(t, \boldsymbol{x}_n) = \frac{1}{3}\sum_{i=1}^{3}\left(\frac{\mathcal{G}_{i4}(t, \boldsymbol{x}_n)}{\mathcal{G}_{i4}^{\rm lat}(t, \boldsymbol{x}_n)}\right)^2. \tag{B.9}$$

In Eq. (B.9), the average over i is taken because generally the ratio $\mathcal{G}_{i4}(t, \boldsymbol{x}_n)/\mathcal{G}_{i4}^{\rm lat}(t, \boldsymbol{x}_n)$ at a lattice site $\boldsymbol{x}_n$ depends on i. However, in our particular choice of discretization, i.e., the Wilson gauge actions and the clover-leaf operator, it is easily shown that $\mathcal{G}_{i4}(t, \boldsymbol{x}_n)/\mathcal{G}_{i4}^{\rm lat}(t, \boldsymbol{x}_n)$ does not depend on i. In this special case, the average over i in Eq. (B.9) is redundant. When this property is violated, the improvement of $\langle U_{\gamma\gamma'}(t, \boldsymbol{x}_n)\rangle_Q$ may be replaced by the one that depends on γ and γ' in place of Eq. (B.7).

There is one more subtle issue about Eq. (B.7). At the tree level, one easily finds that the matrix elements of U_{44}, U_{rr} and $-U_{\theta\theta}$ are the same, while the actual lattice data do not necessarily satisfy such a relation as shown in the main text. In our tree-level improvement, therefore, we decompose our lattice data into the treelike part and the rest, $\langle U_{\gamma\gamma'}(t, \boldsymbol{x}_n)\rangle_Q = U^{\rm tl}_{\gamma\gamma'}(t, \boldsymbol{x}_n) + \delta U_{\gamma\gamma'}(t, \boldsymbol{x}_n)$, where the treelike part $U^{\rm tl}_{\gamma\gamma'}(t, \boldsymbol{x}_n)$ satisfies $U^{\rm tl}_{44}(t, \boldsymbol{x}_n) = U^{\rm tl}_{rr}(t, \boldsymbol{x}_n) = -U^{\rm tl}_{\theta\theta}(t, \boldsymbol{x}_n)$. Then, we apply our tree-level improvement only to the first term:

$$\langle U_{\gamma\gamma'}(t, \boldsymbol{x}_n)\rangle_Q^{\mathrm{imp}} = c(t, \boldsymbol{x}_n)U^{\mathrm{tl}}_{\gamma\gamma'}(t, \boldsymbol{x}_n) + \delta U_{\gamma\gamma'}(t, \boldsymbol{x}_n). \tag{B.10}$$

Below, we consider three choices of $U^{\mathrm{tl}}_{\gamma\gamma'}(t, \boldsymbol{x}_n)$ to estimate the systematic uncertainty of this procedure: $U^{\mathrm{tl}}_{\gamma\gamma'} = \langle U_{44}\rangle_Q$, $\langle U_{rr}\rangle_Q$, and $-\langle U_{\theta\theta}\rangle_Q$. Corresponding results for $\langle \mathcal{T}^{\mathrm{R}}_{rr}(t, r)\rangle$ as example are shown in Fig. B.1(b)–(d), respectively, together with the case without the correction, Fig. B.1(a). Colored open symbols represent the data at each N_τ, and the gray (yellow) shade is the continuum result with (without) the data at $N_\tau = 10$. The figures show that the tree-level improvement suppresses the discretization effect at short distances in three cases, especially for $N_\tau = 10$.

Let us now compare the continuum extrapolation using the data including $N_\tau = 10$ with the tree-level improvement and that using the data without $N_\tau = 10$ and without the tree-level improvement. Shown in Fig. B.2 is such a comparison for $-\langle \mathcal{T}_{rr}(t, rT = 0.40)\rangle_Q/T^4$ at $T/T_c = 1.44$ as functions of $1/N_\tau^2$. Colored open triangles represent the data without the tree-level improvement (red) and with the tree-level improvement for three different prescriptions (blue, orange, and green). Filled symbols at $1/N_\tau^2 = 0$ are the continuum extrapolation: The red squares are continuum results without $N_\tau = 10$ data discussed in the main text, while the diamonds are the continuum results with $N_\tau = 10$ data after the tree-level improvement. Taking into the uncertainty associated with the different prescriptions for the tree-level improvement, the default results without $N_\tau = 10$ in the main text are found to be consistent with the improved results including $N_\tau = 10$.

References

1. Necco S, Sommer R (2002) The N(f) = 0 heavy quark potential from short to intermediate distances. Nucl Phys B 622:328–346. arXiv:hep-lat/0108008 [hep-lat]
2. Kaczmarek O, Karsch F, Zantow F, Petreczky P (2004) Static quark anti-quark free energy and the running coupling at finite temperature. Phys Rev D 70:074505. [erratum: Phys Rev D 72:059903 (2005)]. arXiv:hep-lat/0406036 [hep-lat]
3. Fodor Z, Holland K, Kuti J, Mondal S, Nogradi D, Wong CH (2014) The lattice gradient flow at tree-level and its improvement. JHEP 09:018. arXiv:1406.0827 [hep-lat]
4. Altenkort L, Eller AM, Kaczmarek O, Mazur L, Moore GD, Shu HT (2021) Heavy quark momentum diffusion from the lattice using gradient flow. Phys Rev D 103(1):014511. arXiv:2009.13553 [hep-lat]
5. Fritzsch P, Ramos A (2013) The gradient flow coupling in the Schrödinger Functional. JHEP 10:008. arXiv:1301.4388 [hep-lat]

Appendix C
Leading-Order Perturbative Analysis of EMT around Single Static Charge

In this appendix, we perturbatively calculate the correlation function between the Polyakov loop and the EMT operator in the SU(N) Yang–Mills system at high temperature where $g(2\pi T) \ll 1$. The effective theory valid at the length scale of $R \gg (2\pi T)^{-1}$ is the dimensionally reduced electrostatic QCD (EQCD) in three dimensions (see, e.g., [1–4]),

$$S_{\text{EQCD}} = \int d^3x \left[\frac{1}{2} \text{Tr} \mathcal{G}^2 + \text{Tr}(D\varphi)^2 + m_{\text{D}}^2 \text{Tr}\varphi^2 + \delta\mathcal{L}_{\text{EQCD}} \right]. \tag{C.1}$$

The cardinal contribution to this IR theory comes from the Matsubara zero mode because the other modes exponentially fall off at $R \gg (2\pi T)^{-1}$. Here, $(\mathcal{A}_i, \varphi) = (\mathcal{A}_i^a t^a, \varphi^a t^a) = (A_i, A_4)/(g\sqrt{T})$, $\mathcal{G}_{ij} = \partial_i \mathcal{A}_j - \partial_j \mathcal{A}_i + i g_{\text{E}} [\mathcal{A}_i, \mathcal{A}_j]$, and $D_i \varphi = \partial_i \varphi + i g_{\text{E}} [\mathcal{A}_i, \varphi]$ with $\text{Tr}(t^a t^b) = \delta^{ab}/2$. The higher-dimensional operators are denoted by $\delta\mathcal{L}_{\text{EQCD}}$, which is given by the integration over the nonzero Matsubara modes. The effective coupling and the Debye screening mass in the leading-order (LO) read $g_{\text{E}} = g\sqrt{T}$ and $m_{\text{D}}^2 = (N/3)(gT)^2$, respectively.

Under the 'Feynman static gauge,' which means $\partial_4 \mathcal{A}_4 = 0$ for the temporal component and the Feynman gauge for the spatial component $\mathcal{A}_i$ [2, 3], the tree-level propagators read

$$\langle \varphi^a(\mathbf{0}) \varphi^b(\boldsymbol{x}) \rangle = \delta_{ab} \frac{e^{-m_{\text{D}}|\boldsymbol{x}|}}{4\pi |\boldsymbol{x}|}, \tag{C.2}$$

$$\langle \mathcal{A}_i^a(\mathbf{0}) \mathcal{A}_j^b(\boldsymbol{x}) \rangle = \delta_{ab} \delta_{ij} \frac{1}{4\pi |\boldsymbol{x}|}, \tag{C.3}$$

$$\langle \mathcal{A}_i^a(\mathbf{0}) \varphi^b(\boldsymbol{x}) \rangle = 0, \tag{C.4}$$

where a and b are color indices. Moreover, the Polyakov loop operator $\Omega(\boldsymbol{x})$ is written as

$$\Omega(\boldsymbol{x}) = \mathcal{P} e^{-ig \int_0^{1/T} d\tau A_4(\boldsymbol{x}, \tau)} = e^{-ig\varphi(\boldsymbol{x})/\sqrt{T}}. \tag{C.5}$$

R. Yanagihara, *Distribution of Energy Momentum Tensor around Static Charges in Lattice Simulations and an Effective Model*, Springer Theses,
https://doi.org/10.1007/978-981-16-6234-8

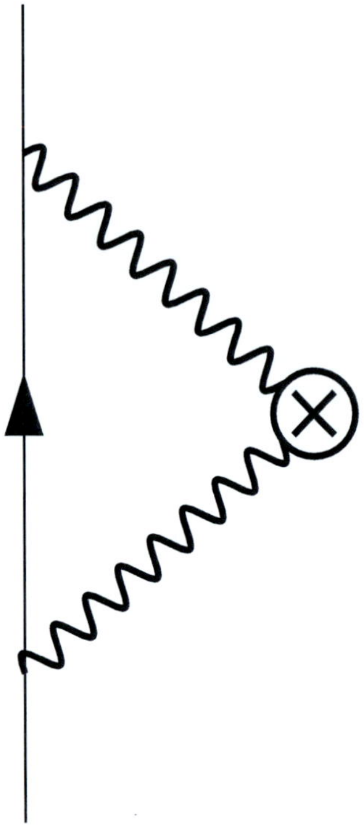

Fig. C.1 Diagram contributing to the leading-order calculation of the correlation function between the Polyakov loop and the EMT operator. The vertical line represents the Polyakov loop and two wavy lines exchanged gluons. The symbol ⊗ corresponds to the EMT operator

The LO contribution to the connected correlation between the Polyakov loop and the EMT stems from the two-gluon exchange of $O(g^2)$ and is diagrammatically shown in Fig. C.1. Since the Polyakov loop operator has only the scalar component $\varphi(\boldsymbol{x})$, the terms which survive in the LO are the connected diagrams with G_{4i}^2, i.e.,

$$\langle (G_{i4}^a)^2(\boldsymbol{x})\rangle_Q = \frac{\langle (G_{i4}^a)^2(\boldsymbol{x})\mathrm{Tr}\Omega(\boldsymbol{0})\rangle_c}{\langle \mathrm{Tr}\Omega(\boldsymbol{0})\rangle}, \tag{C.6}$$

where the suffix c implies the connect correlation.

By expanding $(G_{i4}^a)^2$ and $\mathrm{Tr}\Omega$ up to $O(\varphi^2)$ for fixed $i = 1, 2, 3$, we obtain

$$\begin{aligned}
&\langle (G_{i4}^a)^2(\boldsymbol{x})\rangle_Q \\
&= -\frac{1}{4N} g^2 \langle \varphi^a(\boldsymbol{0})\partial_i\varphi^b(\boldsymbol{x})\rangle\langle \varphi^b(\boldsymbol{0})\partial_i\varphi^a(\boldsymbol{x})\rangle + O(g^3) \\
&= -\frac{N^2-1}{4N} g^2 \left\{\partial_i\left(\frac{e^{-m_{\mathrm{D}}|\boldsymbol{x}|}}{4\pi|\boldsymbol{x}|}\right)\right\}^2 + O(g^3) \\
&= -\frac{C_F}{8\pi}\alpha_s \frac{x_i^2(m_{\mathrm{D}}|\boldsymbol{x}|+1)^2}{|\boldsymbol{x}|^6} e^{-2m_{\mathrm{D}}|\boldsymbol{x}|} + O(g^3),
\end{aligned} \tag{C.7}$$

where $\alpha_s = g^2/4\pi$ and $C_F = (N^2-1)/2N$.

Picking up the contributions of $(G^a_{4i})^2(\boldsymbol{x})$ in each component of the EMT, we obtain the following perturbative estimate for $r \equiv |\boldsymbol{x}| \gg (2\pi T)^{-1}$ up to $O(g^2)$,

$$\langle \mathcal{T}_{44}(\boldsymbol{x})\rangle_Q = \langle \mathcal{T}_{rr}(\boldsymbol{x})\rangle_Q = -\langle \mathcal{T}_{\theta\theta}(\boldsymbol{x})\rangle_Q = -\frac{C_F}{8\pi}\alpha_s \frac{(m_{\mathrm{D}} r + 1)^2}{r^4} e^{-2m_{\mathrm{D}} r} + O(g^3), \tag{C.8}$$

The simplest way to show the above relation is to choose $\boldsymbol{x} = (r, 0, 0)$, so that $\mathcal{T}_{rr}(r, 0, 0) = \mathcal{T}_{11}(r, 0, 0)$ and $\mathcal{T}_{\theta\theta}(r, 0, 0) = \mathcal{T}_{22}(r, 0, 0)$.

Although one finds that the EMT trace $\langle \mathcal{T}_{\mu\mu}(\boldsymbol{x})\rangle_Q$ vanishes at $O(g^2)$, one can utilize the following trace anomaly to evaluate its $O(g^4)$ contribution:

$$\mathcal{T}_{\mu\mu} = \frac{\beta}{2g} G^a_{\mu\nu} G^a_{\mu\nu}, \tag{C.9}$$

where the Yang–Mills beta function reads $\beta = -\beta_0 g^3 - \beta_1 g^5 + \cdots$, with $\beta_0 = (11/3)C_A/(4\pi)^2$, $\beta_1 = (34/3)C_A^2/(4\pi)^4$, and $C_A = N$. By using the right-hand side of the formula and following the same procedure as above, we find

$$\langle \mathcal{T}_{\mu\mu}(\boldsymbol{x})\rangle_Q = -\frac{11}{3}\frac{C_A C_F}{(4\pi)^2}\alpha_s^2 \frac{(m_{\mathrm{D}} r + 1)^2}{r^4} e^{-2m_{\mathrm{D}} r} + O(g^5). \tag{C.10}$$

This is indeed $O(g^2)$ higher than Eq. (C.8).

References

1. Appelquist T, Pisarski RD (1981) High-temperature Yang-Mills theories and three-dimensional quantum chromodynamics. Phys Rev D 23:2305
2. D'Hoker E (1982) Perturbative results on QCD in three-dimensions at finite temperature. Nucl Phys B 201:401–428
3. Nadkarni S (1983) Dimensional reduction in hot QCD. Phys Rev D 27:917
4. Braaten E, Nieto A (1995) Asymptotic behavior of the correlator for Polyakov loops. Phys Rev Lett 74:3530–3533. arXiv:hep-ph/9410218 [hep-ph]

Appendix D
Smearing Method

In this appendix, we briefly introduce the APE smearing method [1], which we employ in Chap. 6 to enhance a coupling of the Wilson loop to the ground state in the $Q\bar{Q}$ system. We comment that the APE smearing is a kind of discrete transformation in contrast to the gradient flow, which is a continuous one we also use in our study to define the EMT operator.

The purpose of the smearing is to reduce the ultraviolet (UV) fluctuations of quantum field configurations, while the IR structure is unchanged. In the smearing procedure on the lattice, we substitute a weighted average link for an original link. We refer to the former (latter) variable as a 'fat' ('thin') link. Since the smearing suppresses UV fluctuations, we can reduce the noise in numerical simulations, leading to measurement of, for example, the topological charge.

However, the smearing may, at the same time, cause a deviation out of the gauge group under consideration. Thus, we have to perform a projection of the smeared variables into the original gauge group at every smearing procedures. For the projection, a kind of maximization procedure is often employed.

The summary of the smearing process is as follows:

1. We replace a thin link variable by a new fat link with the neighboring link variables.
2. We unitarize the new link variable, which is the projection of the smeared link into the original gauge group.

In the following, we concentrate on the SU(N) gauge group. We first discuss the SU(N) projection and then mention the APE smearing procedure.

D.1 SU(N) Projection

For simplicity, we start with the SU(2) projection. Since a linear combination of elements belonging to the SU(2) gauge group is also a SU(2) element, a new link variable $\tilde{U}'_\mu(n)$ is represented with $\bar{U}_\mu(n)$ as

R. Yanagihara, *Distribution of Energy Momentum Tensor around Static Charges in Lattice Simulations and an Effective Model*, Springer Theses,
https://doi.org/10.1007/978-981-16-6234-8

$$\tilde{U}'_\mu(n) = k\bar{U}'_\mu(n), \tag{D.1}$$

where

$$k^2 = \det\left(\tilde{U}'_\mu(n)\right). \tag{D.2}$$

When $\mathrm{Re\,Tr}[U_\mu(n)\tilde{U}'_\mu(n)]$ is maximized, $\mathrm{Re\,Tr}[1 - U_\mu(n)\tilde{U}'_\mu(n)]$ is minimized. In particular, $\mathrm{Re\,Tr}[U_\mu(n)\tilde{U}'_\mu(n)] = 1$ indicates the unitarity. The maximization is achieved in the condition,

$$\mathrm{Re\,Tr}[U_\mu(n)\tilde{U}'_\mu(n)] = \mathrm{Re\,Tr}[1]. \tag{D.3}$$

Therefore, we obtain

$$\bar{U}_\mu(n)^{-1} = \bar{U}_\mu(n)^\dagger = \frac{(\tilde{U}'_\mu(n))^\dagger}{k}. \tag{D.4}$$

Now, we proceed to the case of the SU(3) gauge group. A sum of elements in the SU(3) gauge group, however, is not included in the original group generally. In the case of the SU(3) gauge group, we have to iteratively perform the above projection for three SU(2) subgroups of the SU(3) group while wrapping the whole SU(3) group. Thus, the SU(3) projection of a new link variable $\tilde{U}'_\mu(n)$ is defined by

$$\mathrm{Proj}_{\mathrm{SU(3)}}(\tilde{U}'_\mu(n)) = \max_{U_\mu(n)\in \mathrm{SU(3)}} \mathrm{Re\,Tr}(U_\mu(n)\tilde{U}'_\mu(n)^\dagger). \tag{D.5}$$

What we have to do for the SU(3) projection is to find out $U_\mu(n)$ satisfying the relation Eq. (D.5).

There is another method of the projection such as

$$U = \tilde{U}'(\tilde{U}'^\dagger\tilde{U}')^{-1/2}\det(\tilde{U}'^{-1}\tilde{U}'^\dagger)^{1/6}. \tag{D.6}$$

Since any elements $\tilde{U}'$ satisfy the hermiticity of $\tilde{U}'^\dagger\tilde{U}'$, they are diagonalizable. If $\det(\tilde{U}') \neq 0$, then we can calculate

$$W = \tilde{U}'\frac{1}{\sqrt{\tilde{U}'^\dagger\tilde{U}'}}. \tag{D.7}$$

With the variable W, we can construct

$$U = W\frac{1}{\sqrt[3]{\det W}}, \tag{D.8}$$

which is the projection matrix.

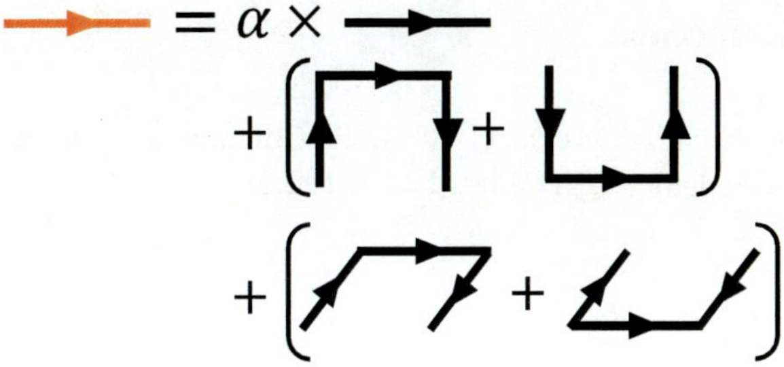

Fig. D.1 Spatial APE smearing. An original thin link is replaced by fat link (orange arrow) with neighboring staples

D.2 APE Smearing

One of the most famous smearing methods is the APE smearing proposed in Ref. [1]. In the APE smearing, we substitute a weighted average link with neighboring staples for an original link as

$$\tilde{U}'_\mu(n) = \alpha U_\mu(n) + \sum_{\mu \neq \nu} C_{\mu\nu}(n), \tag{D.9}$$

where

$$C_{\mu\nu} = U_\nu(n)U_\mu(n+\hat{\nu})U_\nu^\dagger(n+\hat{\mu}) + U_\nu^\dagger(n-\hat{\nu})U_\mu(n-\hat{\nu})U_\nu(n-\hat{\nu}+\hat{\mu}) \tag{D.10}$$

represents the staple and α is a smearing parameter. In particular, the APE smearing only for the spatial links is defined as

$$\tilde{U}'_i(n) = \alpha U_i(n) + \sum_{j \neq i} C_{ij}(n). \tag{D.11}$$

We show Eq. (D.11) diagrammatically in Fig. D.1. The following SU(3) projection is implemented by

$$\mathrm{Proj}_{\mathrm{SU(3)}}(\tilde{U}'_\mu(n)) = \max \mathrm{Re}\,\mathrm{Tr}\big(U_\mu(n)\big[\alpha U_\mu(n)^\dagger + \sum_{\mu \neq \nu} C_{\mu\nu}^\dagger(n)\big]\big). \tag{D.12}$$

In our study discussed in Chap. 6, we employ such a spatial APE smearing for the spatial connecting links in the Wilson loop.

Reference

1. Albanese M et al. (1987) [APE Collaboration], Glueball masses and string tension in lattice QCD. Phys Lett B 192:163

Appendix E
Analytic Properties of Abelian-Higgs Model

In this appendix, we summarize the analytic properties of the magnetic vortex with an infinite length in the AH model at the Bogomol'nyi bound discussed in Ref. [1]. Throughout this appendix, we use the dimensionless variables ρ, $P(\rho)$, and $Q(\rho)$ defined in Eq. (7.39).

The classical vortex solution corresponds to the minimum of Eq. (7.46),

$$\hat{\Sigma}[P, Q] = \int_0^\infty d\rho\, \rho \Big[\frac{1}{2\rho^2} (\partial_\rho Q)^2 + (\partial_\rho P)^2 + \frac{P^2 Q^2}{\rho^2} + \kappa^2 (P^2 - 1)^2 \Big], \tag{E.1}$$

with the boundary conditions,

$$Q(\rho) \to n, \quad P(\rho) \to 0 \quad \text{for } \rho \to 0, \tag{E.2}$$

$$Q(\rho) \to 0, \quad P(\rho) \to 1 \quad \text{for } \rho \to \infty, \tag{E.3}$$

where we allow for an arbitrary winding number n in this appendix.

At $\kappa = 1/\sqrt{2}$, Eq. (E.1) is rewritten as

$$\hat{\Sigma}[P, Q] = \int_0^\infty d\rho\, \rho \Big[\frac{1}{2} A_\pm(\rho)^2 + B_\pm(\rho)^2 \Big] \mp \int_0^\infty d\rho\, \partial_\rho (Q(P^2 - 1)), \tag{E.4}$$

with

$$A_\pm(\rho) = \frac{\partial_\rho Q}{\rho} \pm (P^2 - 1), \quad B_\pm(\rho) = \partial_\rho P \pm \frac{PQ}{\rho}. \tag{E.5}$$

Using these variables, the dimensionless EMT $\hat{\mathcal{T}}_{rr}(\rho)$ and $\hat{\mathcal{T}}_{\theta\theta}(\rho)$ in Eq. (7.40) are given by

R. Yanagihara, *Distribution of Energy Momentum Tensor around Static Charges in Lattice Simulations and an Effective Model*, Springer Theses,
https://doi.org/10.1007/978-981-16-6234-8

$$\hat{\mathcal{T}}_{rr}(\rho) = \frac{1}{4}A_+(\rho)A_-(\rho) + \frac{1}{2}B_+(\rho)B_-(\rho), \tag{E.6}$$

$$\hat{\mathcal{T}}_{\theta\theta}(\rho) = -\frac{1}{4}A_+(\rho)A_-(\rho) + \frac{1}{2}B_+(\rho)B_-(\rho). \tag{E.7}$$

The last term in Eq. (E.4) given by the total derivative is calculated to be

$$\mp \int_0^\infty d\rho\, \partial_\rho (Q(P^2 - 1)) = \mp n, \tag{E.8}$$

with the boundary conditions, Eqs. (E.2) and (E.3), and one obtains

$$\hat{\Sigma}[P, Q] = \int_0^\infty d\rho\, \rho \Big[\frac{1}{2}A_\pm(\rho)^2 + B_\pm(\rho)^2\Big] \mp n. \tag{E.9}$$

Then, the minimum of Eq. (E.9) is obtained when $A_\pm(\rho) = 0$ and $B_\pm(\rho) = 0$ are satisfied for one of the signs of subscript [1]. Assuming that $P(\rho)$ and $Q(\rho)$ are monotonic functions of ρ in the vortex solution, from the boundary conditions one finds that the conditions $A_+(\rho) = 0$ and $B_+(\rho) = 0$ are excluded for positive n, and hence the vortex solution must satisfy

$$A_-(\rho) = B_-(\rho) = 0. \tag{E.10}$$

Substituting Eq. (E.10) into Eq. (E.9), one obtains $\hat{\sigma}_{\mathrm{AH}}(1/\sqrt{2}) = n$ [1]. For $n < 0$, $A_+(\rho) = 0$ and $B_+(\rho) = 0$ are satisfied, and for an arbitrary n one obtains

$$\hat{\sigma}_{\mathrm{AH}}(\kappa) = |n|. \tag{E.11}$$

at $\kappa = 1/\sqrt{2}$.

Finally, substituting Eq. (E.10) into Eqs. (E.6) and (E.7), one easily finds that $\mathcal{T}_{rr}(r) = \mathcal{T}_{\theta\theta}(r) = 0$ at $\kappa = 1/\sqrt{2}$.

Reference

1. Bogomol'nyi EB (1976) The stability of classical solutions. Sov J Nucl Phys 24:449 PRINT-76-0543 (LANDAU-INST)